Map of the World

An Introduction to Mathematical Geodesy

Map of the World

An Introduction to Mathematical Geodesy

Martin Vermeer
Antti Rasila

CRC Press
Taylor & Francis Group
Boca Raton London New York

CRC Press is an imprint of the
Taylor & Francis Group, an **informa** business

CRC Press
Taylor & Francis Group
6000 Broken Sound Parkway NW, Suite 300
Boca Raton, FL 33487-2742

First issued in paperback 2021

© 2020 by Taylor & Francis Group, LLC
CRC Press is an imprint of Taylor & Francis Group, an Informa business

No claim to original U.S. Government works

ISBN-13: 978-0-367-21773-0 (hbk)
ISBN-13: 978-1-03-217754-0 (pbk)
DOI: 10.1201/9780429265990

Publisher's Note

The publisher has gone to great lengths to ensure the quality of this reprint but points out that some imperfections in the original copies may be apparent.

**Visit the Taylor & Francis Web site at
http://www.taylorandfrancis.com**

**and the CRC Press Web site at
http://www.crcpress.com**

Contents

8 The geometry of the ellipsoid of revolution 91

9 Three-dimensional co-ordinates and transformations 109

10 Co-ordinate reference systems 127

11 Co-ordinates of heaven and Earth — 143

12 The orbital motion of satellites — 161

13 The surface theory of Gauss — 181

Preface

Travelling from place to place is a basic condition of life for humans as it is for animals. Although sometimes the journey may be more important than the destination, one generally needs guidance in travelling, e.g., by landmarks. Many animals, such as migrating birds and bees, are able to use landmarks on a level that mankind has, through technology, only reached during the past century. The *map* is an essential means of storing and mediating this vital knowledge.

Nearly all human activity requires knowledge of the living environment. Maps, geospatial data in graphical form, are used for travelling, like seafaring, aviation and road transport, national defence, safety and security, as well as tasks related to, e.g., rescue operations. Map-based systems maintain the smooth functioning of society, because knowledge of both the borders of states and those of land parcels, as well as of land ownership, must be recorded precisely and reliably. Geospatial information and maps are needed also for construction, and, e.g., for weather prediction. For these reasons, they are a precondition for economic development.

Thus, maps are essential to visualization, commerce, navigation, and positioning services. The map is an important source of background information for presenting and optimizing routes for aviation, seafaring and other means of transport. Technological development has given us besides traditional maps, also mobile technology and satellite positioning. They are already almost indispensable tools for using map materials.

Nowadays a map is not just an image of the terrain, but rather a digital data source containing all sorts of knowledge. Thanks to wireless communication, map and location services are being used more than ever in real time. Map software then has to be compatible, e.g., with various terminal devices.

Maps and geospatial data must therefore be of good quality and up to date. They should be compatible both nationally and internationally, and both location data and attribute data must be *valid*. The optimal use of map information, however, demands from the user an understanding of the processes by which this information is produced.

This book aims to present the most pertinent mathematical foundations relating to maps, mapping and the production of geospatial data. It is aimed at students of land surveying and natural science as well as professionals, but also at map amateurs. These foundations have a very broad historical context, because historically, mapping the environment we live in is part of a much more ambitious goal. Making maps is part of the human endeavour to understand what is our place in the universe. Throughout history, mapping our planet has gone hand in hand with the rise of science. It connects to so many fundamental questions of science: how does the Earth move in space and time? What is her precise figure? What is the precise form and historical origin of spacetime itself? The map of the world is in itself a great scientific adventure.

Acknowledgments

This book is based on the Finnish edition, *Maailman kartta*, published in 2014 by Ursa Astronomical Association, Helsinki.

Thanks are due especially to Ursa's editorial team, who, in two rounds of extensive critical remarks and correction proposals, helped to improve materially the quality of the manuscript. Constructive proposals were given also by colleagues, students, friends and family members. We mention especially docent Matti Lehtinen, Professor Matti Vuorinen, university lecturer Jarmo Malinen, lecturer emeritus Simo K. Kivelä, Mrs. Ritva Rasila and Mrs. Liisa Vermeer. The final text, however, remains the responsibility of us, the authors.

Financial support for writing the manuscript by *Maanmittausalan edistämis-säätiö*, the Finnish Foundation for the Surveying Profession, is gratefully acknowledged.

After publication, an extensive list of corrections was received from Professor Jan Westerholm of the High Performance Computing laboratory at Åbo Akademi University in Turku. He went carefully through the formulas and derivations of the book using both symbolic algebra and numerical tools. This deserves a big thank you!

Acronyms

ANNA-1B	"Army, Navy, NASA, Air Force," a satellite carrying flashlights, launched in 1962
BIH	*Bureau International de l'Heure*, International Time Bureau
CIO	conventional international origin (of polar motion)
CTS	Conventional Terrestrial System
DCF77	time-signal radio transmitter near Frankfurt a. M., Germany
ECEF	Earth-centred, Earth-fixed
ED50	European Datum 1950
ee	equation of equinoxes
EOP	Earth Orientation Parameters
ETRF	European Terrestrial Reference Frame
ETRS	European Terrestrial Reference System
ETRS-GKn	Finnish Gauss–Krüger based projected co-ordinates in the EUREF-FIN datum, used for large-scale maps. The integer n is the Eastern longitude of the central meridian
ETRS-TM35FIN	Finnish UTM-based (zone 35) projected co-ordinates in the EUREF-FIN datum, used for small-scale maps
EU	European Union
EUREF-FIN	Finnish national ETRS-based reference frame
FinnRef	Finnish continuously operating GNSS reference network
GAST	Greenwich Apparent Sidereal Time
GLONASS	Russian global satellite positioning system
GMST	Greenwich Mean Sidereal Time
GMT	Generic Mapping Tools
GNSS	Global Navigation Satellite Systems
GPS	US Global Positioning System
GRS80	Geodetic Reference System 1980
IAG	International Association of Geodesy
ICRF	International Celestial Reference Frame
ICRS	International Celestial Reference System
IERS	International Earth Rotation and Reference Systems Service
IGS	International GNSS Service

IGY	International Geophysical Year
ISO	International Organization for Standardization
ITRF	International Terrestrial Reference Frame
ITRS	International Terrestrial Reference System
ITS	instantaneous terrestrial system
IUGG	International Union of Geodesy and Geophysics
JHS	(Finnish) Recommendation for Public Administration
JPL	(NASA) Jet Propulsion Laboratory
KKJ	Finnish National Map Grid Co-ordinate System (obsolete)
LAST	Local Apparent Sidereal Time
LCC	Lambert Conformal Conical (map projection)
LMST	Local Mean Sidereal Time
LoD	Length of day
MATLAB®	Matrix Laboratory. MATLAB is a registered trademark of The MathWorks, Inc.[1]
NAD 83	North American Datum 1983
NEU	North, East, Up (co-ordinates)
NTP	Network Time Protocol
ODE	Ordinary Differential Equation
Pageos	Passive Geodetic Satellite, launched in 1966
RAS	real astronomical system
SVD	Singular-Value Decomposition
UNESCO	United Nations Educational, Scientific and Cultural Organization
UTC	Universal Time Co-ordinated
UTM	Universal Transverse Mercator (map projection)
VLBI	very long baseline interferometry
VVJ	Finnish "Old State System," KKJ precursor (obsolete)
WGS84	World Geodetic System 1984
YKJ	Finnish Uniform Co-ordinate System (obsolete)

---

[1] 3 Apple Hill Drive, Natick MA 0176-2098 USA, Tel. 508-647-7000, Fax 508-647-7001, Email: mailto:info@mathworks.com, Web: https://www.mathworks.com

Chapter 1

A brief history of mapping

The ancient Greek knew that the Earth is a sphere. They travelled long distances and wrote detailed travel accounts. Their understanding of the size of the Earth was however fuzzy. Only much later, at the dawn of the Renaissance, narratives started to give way to measurements. Thus, a better understanding started to form of the Earth's true dimensions. At the same time arose a need, a drive to augment the tales from far-away lands with mathematical models based on geography, as well as collect quantitative data on the Earth and the world more generally.

The early Renaissance was the era when, for the first time, geometrically correct maps of the whole world were produced. Drafting these was complicated by the target being on a curved surface, and thus a direct depiction onto the map plane as a scaled image is impossible. The art of book printing, developed in the 15th century, made the dissemination of documents a lot easier than before. The voyages of discovery and the subsequent military and commercial seafaring — often the same thing — made the drafting of sufficiently precise maps a profitable commercial activity.

Of the map makers of the era, the most famous are Gerardus Mercator, Willem Janszoon Blaeuw, Abraham Ortelius (Ortels) and Jodocus Hondius (Joost de Hondt). It was a time of blossoming of science, erudition and craftsmanship. Also geodetic measurement instruments were developed, and in the Netherlands, Willebrord Snellius[1], a student of Frisius[2] (Haasbroek, 1968), executed an extensive triangulation to determine the true size of the Earth. This was a time of change, but also — or perhaps just because of the changes, which were experienced as traumatic — an age of fanatical religiosity and dark superstition.

At the same time, also the true size of the Earth started to become evident. A little later, the Earth started to be studied as one planet among others. Studies were undertaken on how the Earth and other celestial bodies move in their orbits, like parts of a celestial clockwork. Also this research was, back

---

[1]Willebrord Snel van Royen (1580 – 1626) was a Dutch astronomer, physicist and mathematician.

[2]Gemma Frisius (1508 – 1555) was a Frisian universal genius, physician, mathematician, geodesist and builder of measurement instruments.

then, a pre-condition to, and an enabler of, seafaring. This was the beginning of an era of ever more precise measurement and mapping, which continues to this day.

1.1   Map and scale

Figure 1.1: Gerardus Mercator, engraving Nicolas III de Larmessin 1682.

Ideally, a map of the world would be isometric. An isometric map displays the terrain on a reduced scale. Distances measured on the map correspond, directly through the scale, to distances in the terrain. The idea of an isometric map works in mapping a small area (i.e., on a large scale[3]), but, due to the curvature of the Earth, it is impossible to draft such a plane map for a larger area. Because globes are impractical, we will in this book concentrate on plane maps.

The impossibility of an isometric map can be visualized by thinking of a circle drawn on a sphere, and its radius. We assume here the centre to be also on the surface of the sphere. In the plane, the ratio between the radius of a circle and the length of its circumference is 2π. Because of the assumed isometricity of the map, the same should also be true on the surface of the sphere, when we take the distance between two points to mean the length of the shortest path between them on the surface of the sphere. On the surface of the sphere, however, the ratio is always a different one: the magnitude of the distortion depends on the radius of the circle. Therefore, a map drawn on the plane will always depict part of the surface of the sphere in some map projection that distorts sizes, though from the concept of map scale that is used for maps, one could infer otherwise. The magnitude of the error depends on the extent of the area depicted in the map. In some map projections, like the Mercator projection, the poles are located at an infinite distance from any other point.

---

[3]1:1000,000 is a small, 1:1000 a large scale!

1.2 Gerardus Mercator and Mercator's map of the world

Gerard Kremer (1512–1594) was born in Rupelmonde near Antwerp. His parents were German merchants, who had fled religious war and persecution of protestants in Flanders. Like many scholars in those days, he latinized his name, and it is better known as Gerardus Mercator, see Figure 1.1.

Mercator studied at the University of Leuven, as a student of the famous humanist Macropedius[4]. He was, however, disappointed with the teachings of theology and philosophy, which offered explanations that, in his opinion, did not satisfactorily describe the world and contradicted each other. Mercator traveled widely, returning however back to Leuven, this time as a student of the leading mathematician and astronomer of the time, Gemma Frisius. Frisius had other famous students, like the astronomer Johannes Stadius[5] and advisor to Queen Elisabeth I of England, John Dee[6]. (A crater on the Moon is named after Frisius, and another one after Mercator.)

Mercator became familiar with map making while working as a student of Frisius, who had made many significant inventions relating to this field, such as triangulation (Figure 1.2) and a method to determine the meridian when the clock time is precisely known (Sobel, 1995). The first map that was independently prepared by Mercator was from 1537 and was a map of Palestine. After this came a world map (1538) and a map of Flanders (1540). Figure 1.3 shows Mercator's map of the world of 1569.

In 1544 Mercator was arrested and convicted of heresy. Reasons for this may have been the suspicions aroused by his many information-gathering journeys, as well as his protestant views. The sentence was seven months of imprisonment, which he served in the castle of Rupelmonde.

In 1554 Mercator moved to Duisburg, Germany. It has been surmised that he left the Netherlands for religious reasons, or having heard of a plan to found a new university. Mercator taught mathematics at the University of Duisburg while developing a new map projection. In 1569 he published his most important work, the famous world map *Nova et Aucta Orbis Terrae Descriptio ad Usum Navigatium Emendate, "New and precise presentation of the globe for use in navigation at sea."* Mercator also took into use the term

---

[4]Georgius Macropedius (1487–1558) was a Dutch humanist and school master. He is considered the most important author of plays in the Latin language of the 16th century.

[5]Johannes Stadius, or Estadius, (1527–1579) was a Flemish astronomer and mathematician.

[6]John Dee (1527–1608 or 1609) was a Welsh mathematician, astronomer, astrologer, occultist, navigator, and advisor to Queen Elisabeth I. He dedicated a large part of his life to alchemy, fortune telling, and studying hermeticism. John Dee is also known for the code 007 used in espionage fiction, with which he signed letters meant for Her Majesty's eyes only.

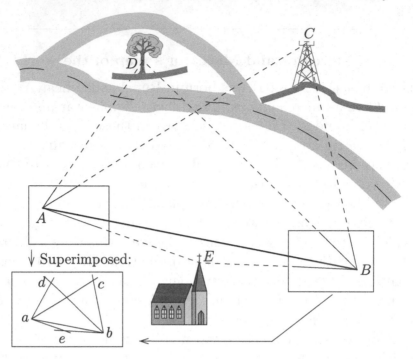

Figure 1.2: This is how triangulation works: in two points A and B we draw, using a so-called plane-table alidade, directions to various objects in the terrain onto paler, and later in the office, the papers are superimposed, forming a map. The *scale* of the whole map is obtained by measuring the true distance AB in the terrain, and the corresponding distance ab on the map. This simple method for mapping the terrain using a plane table and alidade was apparently still used in the 1980s.

atlas, meaning a uniformly edited map publication consisting of many map sheets, which together form a presentation of a surface in three-dimensional space. Although the word nowadays refers to any kind of book of maps, it is used in nearly its original meaning in the theory of Riemann surfaces and more generally in differential geometry.

The story of the life and work of Gerardus Mercator is told expertly in a fairly new biography (Crane, 2002), which describes in a lively fashion the spirit of the era. Clearly Mercator's world was very different from our own. However, the attentive reader will notice one frightening similarity. Although, in today's world, being burned at the stake is no longer available as a criminal punishment option, the public at large, and politicians, do not always hold science or scholarship in any higher regard than was the case in Mercator's days (Mann, 2012).

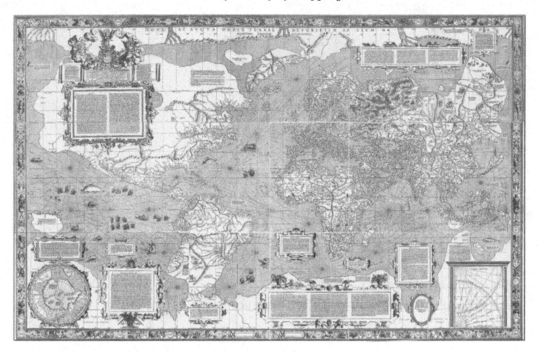

Figure 1.3: Mercator's map of the world of 1569 (Wikipedia, Mercator 1569 world map).

1.3   Construction of the Mercator projection

The Mercator projection is *conformal*: it preserves the angle between two curves at their point of intersection. It is the only conformal cylinder projection. This property is useful in a map projection, because then, measurements made in the map correspond to directions measured in the terrain. Because the meridian and the parallel can be determined by measuring the elevation angles of celestial bodies at known times, and the direction is obtained with a compass, it is extremely well suited for navigation.

Mercator did not himself present any mathematical explanation for his map projection. In 1599 Edward Wright[7] discovered how the projection can be presented mathematically. Consider a small area, the Southern bound of which is on the parallel ϕ. The bounds of the area are along meridians and parallels and the lengths of both the Western and the Southern boundary are h. In order for the Mercator projection to work, the shape of the area on the map must be a square. Let us designate the meridian of the area's Western boundary by λ.

---

[7]Edward Wright (1561–1615) was an English mathematician, cartographer, navigator at sea (and pirate in the service of the state), instrument maker and land surveyor.

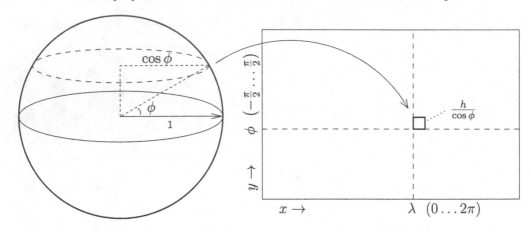

Figure 1.4: Construction of Mercator's map projection.

Let us fix the image point (x, y) and choose $x = \lambda$. We are left with the task to calculate y. Geometrically, we may infer (see Figure 1.4) that the map-projection stretch for latitude ϕ is $\dfrac{1}{\cos \phi}$. Thus, the area the width of which is h, has a width in the map of $\dfrac{h}{\cos \phi}$. Therefore, also the height has to be $\dfrac{h}{\cos \phi}$.

The y co-ordinate of the image point depends only on the latitude ϕ. We may therefore write $y = F(\phi)$, in which F is a strictly increasing function. We must find out what F is. From the image of the area on the map, we obtain an equation, that can be written using F:

$$F(\phi + h) = F(\phi) + \frac{h}{\cos \phi},$$

i.e.,

$$\frac{F(\phi + h) - F(\phi)}{h} = \frac{1}{\cos \phi}.$$

If we let $h \to 0$, we obtain as the limit the *derivative*

$$F'(\phi) = \frac{1}{\cos \phi}.$$

By fixing the image of the equator in the map to level $y = 0$, we obtain for the y co-ordinate the integral equation

$$F(\phi) = \int_0^\phi \frac{dt}{\cos(t)}. \tag{1.1}$$

Unfortunately, the geometric construction was done before differential and integral calculus, and one was unable to calculate integral 1.1, although today it belongs to high-school math. Approximate values for the integral were

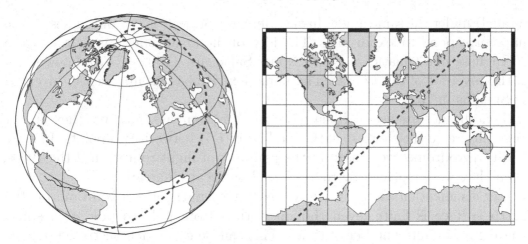

Figure 1.5: The loxodrome or rhumb line resembles on a globe a spiral, but its image on a Mercator map is straight. The loxodrome does not, however, give the shortest path.

published as tables to be used in navigation. John Napier[8] discovered in 1614 the logarithm function, and published in 1620 a table book containing the logarithms of trigonometric functions. When studying such tabular works, Henry Bond[9] noticed by luck in 1640, that

$$\int_0^\phi \frac{dt}{\cos(t)} = \ln\left(\tan\left(\frac{\phi}{2} + \frac{\pi}{4}\right)\right).$$

Proving this result was however left as an open problem until 1668, when James Gregory[10] published for it a (very complicated) proof (Rickey and Tuchinsky, 1980).

The Mercator projection may thus be defined by the formula

$$(x, y) = \left(\lambda, \ln\left(\tan\left(\frac{\phi}{2} + \frac{\pi}{4}\right)\right)\right),$$

in which ϕ is the latitude of the point on the surface of the sphere, and λ its longitude.

1.4 Loxodromes or rhumb lines

A loxodrome or rhumb line is a curve that is described by moving along a fixed compass direction. The loxodrome intersects every meridian (and every

---

[8] John Napier of Merchiston (1550 – 1617) was a Scottish nobleman and mathematician.

[9] Henry Bond (approx. 1600 – 1678) was a teacher of navigation and surveying in Cambridge.

[10] James Gregory (1638 – 1675) was a Scottish astronomer and mathematician.

parallel) under the same angle. In the Mercator projection, straight lines on the map correspond to loxodromes. The idea of the loxodrome was found by the Portuguese mathematician Pedro Nunes Salaciense (1502–1578), known also as the inventor of the *nonius*. The name loxodrome was apparently invented by Snellius in 1624.

Loxodromes are useful when navigating by compass, however they do not (generally) give the shortest path (see Figure 1.5) between two points. In spite of this, loxodromes were used in the planning of flight routes until the 1960s, when they were replaced by routes based on great-circle arcs.

Let us mention, as a curiosity, that the length of a loxodrome on the Earth surface is finite, even though it runs from pole to pole as a spiral winding an infinite number of times. This can be easily demonstrated: let the compass heading of the loxodrome be A (a constant), so its element of length is $ds = \dfrac{d\phi}{\cos A}$, and its total length

$$s = \int_{\text{South Pole}}^{\text{North Pole}} ds = \frac{1}{\cos A} \int_{-\frac{\pi}{2}}^{\frac{\pi}{2}} d\phi = \frac{\pi}{\cos A}.$$

1.5   A political map of the world?

In drawing a map of the world, the objective of Mercator was to create a tool for navigators at sea that was as good as possible. In navigation, it is required that the map projection is conformal, which leads to the loss of other properties. Mercator's projection does not, e.g., preserve surface areas. In the projection, Greenland looks about similar in size to Africa, see Figure 2.5 on page 25. In reality Africa is about 13 times larger in surface area. It follows from the properties of the map, that areas far from the equator are larger on the map than in reality. See Figures 1.6 and 2.5 on page 25.

The German social scientist Arno Peters (1916–2002), in his book *Die Neue Kartographie/The New Cartography* (Peters, 1983), made strong accusations against map makers. According to him, maps present the rich Northern industrial countries as larger than life, and thus more important. The "egalitarian" world map proposed by Peters can be found at Wikipedia, Gall-Peters projection.

To those familiar with the math behind map projections, the criticism of Peters may easily seem over the top. On the other hand, maps certainly shape people's views of the world in many ways (Monmonier, 1996). The usefulness of Mercator's map projection for navigation by compass on the high seas has made it the favourite in many situations for which it is unsuitable. For example, in teaching geography in schools, surely surface areas and distances are more

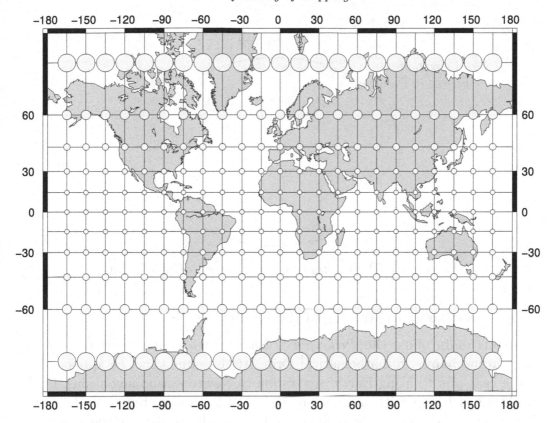

Figure 1.6: Distortion of surface areas in the Mercator projection.

important than the conformality of a map, meaning that a so-called equivalent or equal-area map projection is probably a better choice.

In fact, the map projection proposed by Peters was already invented and published in 1885 by the English vicar James Gall (Gall, 1885) under the name Gall's orthographic cylinder projection. It is certainly not the only equal-area map projection: the cartographic software GMT (Wessel and Smith, 1990), which was used also in creating figures for this book, offers at least ten different equal-area projections, including the Gall–Peters projection.

1.6 Towards the precise figure of the Earth

As science and measurement technologies developed, the notion of the Earth as a mathematical sphere was drawn into question. Theoretical arguments supported the idea that rotational motion would lead to a deviation from the spherical form. Some celestial bodies, like the rapidly rotating planet Jupiter, appear in telescopes clearly flattened.

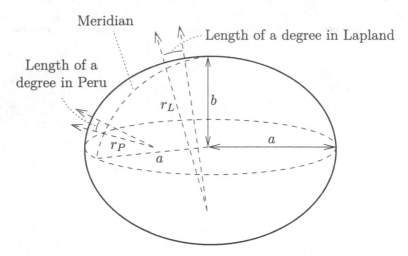

Figure 1.7: Parameters of the ellipsoid of revolution.

1.6.1   Rotation and flattening of the Earth

In his main opus, the *Principia*[11], Isaac Newton presented the laws of classical mechanics in mathematical form. At the same time, he applied them to the case of a rotating Earth. According to Newton, a homogeneous, liquid, equilibrated Earth with a rotation period of 24 hours, between the liquid elements of which acts the force of gravitation, would, because of the centrifugal force, be an ellipsoid of revolution flattened at the poles (Figure 1.7). The definition of flattening, or oblateness, is

$$f = \frac{a - b}{a},$$

in which a and b are the semi-major and semi-minor axes, or equatorial and polar radii. Newton calculated, based on his theory, the value of $f = \frac{1}{230}$ for the flattening.

The idea of a homogeneous-density Earth is just an assumption, and not necessarily a very realistic one: gravity drives the heavy constituents of matter down towards the Earth's centre. The Dutch Christiaan Huygens[12] calculated, assuming all of the Earth's mass to be concentrated in her centre, that the flattening would be $f = \frac{1}{578}$. This result sets a lower bound to the flattening.

As we know today, the truth lies between these extremes. The interior density profile of the Earth is roughly like this (Figure 1.8):

---

[11]*Philosophiæ Naturalis Principia Mathematica* (1687), in modern English, slightly paraphrased, "Mathematical Foundations of Physics."

[12]Christiaan Huygens (1629 – 1695) was a Dutch universal genius, mathematician, astronomer and physicist, who invented, among other things, the pendulum clock, developed the wave theory of light, and recognised the rings of Saturn.

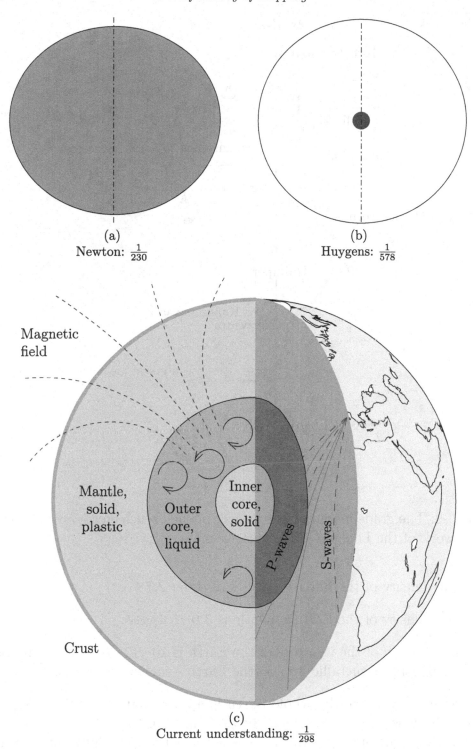

(a)

Newton: $\frac{1}{230}$

(b)

Huygens: $\frac{1}{578}$

Magnetic field

Mantle, solid, plastic

Outer core, liquid

Inner core, solid

P-waves

S-waves

Crust

(c)

Current understanding: $\frac{1}{298}$

Figure 1.8: Different models for the interior mass distribution of the Earth, and the flattening values calculated from them. Newton (left) assumed the mass to be uniformly distributed, whereas Huygens (right) guessed that all mass was in the centre. Our current understanding (below) is between these two: the core is denser than the mantle, and the outer layer is a thin crust.

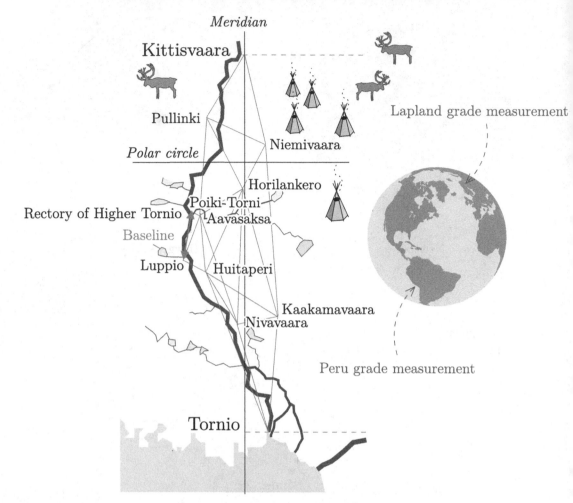

Figure 1.9: The grade measurement project of the French Academy of Sciences: the network of the Lapland grade measurement.

○ The density of the Earth's crust is of order $2.7\,\mathrm{g/cm^3}$,

○ The density of the Earth's mantle is $3.0-5.4\,\mathrm{g/cm^3}$,

○ The density of the iron core of the Earth is $10-13\,\mathrm{g/cm^3}$ (for comparison: the density of metallic iron on the Earth's surface is $7.9\,\mathrm{g/cm^3}$).

The average density of the whole Earth is approximately $5.4\,\mathrm{g/cm^3}$. So, even though the density grows significantly when moving toward the centre of the Earth, a large part of the Earth's mass is nevertheless quite far from her centre. In fact, the radius of the Earth's core is some $3500\,\mathrm{km}$, 55% of the radius of the Earth, and outside it is found some 70% of the mass of the Earth!

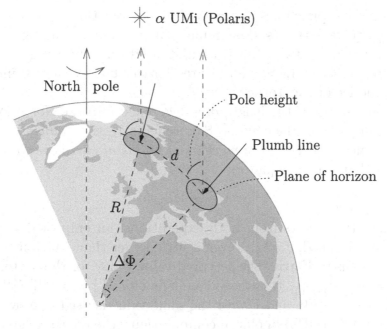

Figure 1.10: Determining the difference $\Delta\Phi$ in directions between plumb lines astronomically. From the difference in direction and the metric distance d, the radius of curvature $R = \frac{d}{\Delta\Phi}$ of the Earth may be calculated.

1.6.2 The big project of the French Academy of Sciences

The problem of the flattening remained unresolved, until half a century later the French Academy of Sciences organized two expeditions. One of the expeditions travelled to Finnish Lapland (1736–37), and the other to South America, Peru (1735–44). The mission of the expeditions was to measure, using astronomical and geodetic measurement techniques, the length of a meridian arc of one degree on two different latitudes, of which one was located close to the equator in Peru, and the other close to the North Pole in Lapland, in the Torne river valley. This was thus a similar grade measurement as the one carried out by the Dutch Snellius over a century earlier.

The basic idea of a grade measurement is seen in Figures 1.7 and 1.10. A baseline is set up in the North-South direction, in the end points of which the directions of the plumb lines differ by about one degree; the plumb-line directions at the end points are measured astronomically.

On the Earth's surface, the horizontal distance between the points is measured in units of length. This all on its own was a substantial part of the work: between the end points, a triangle network was built, and from every point in the network — usually on a mountain top — the directions to neighbouring points were measured, see Figure 1.9. If, in the triangle network, only directions are measured, one may compute precisely the geometry of the network using the principles of *triangulation*; only the absolute size of the network, or, equivalently, its *scale*, remain indeterminate. For this, one of the sides in the triangle network must be measured directly using measurement rods. The Frenchmen in Lapland measured for this purpose a baseline of length 14.5 km on the ice of the Torne river.

If Newton was right, the length of a degree close to the North Pole should be greater than close to the equator. Said differently, the *radius of curvature* of the Earth — more precisely, the *meridional radius of curvature M* — is longer on the poles than on the equator:

$$M_L > M_P.$$

By combining the results from both expeditions, an empirical value for the flattening was obtained, $f = \frac{1}{210}$, even a little larger than Newton's value for a homogeneous-density Earth. The accepted value today for the flattening of the Earth is $f = \frac{1}{298.257\,222\,101}$. It is the value made official by the GRS80 (Geodetic Reference System 1980), and *almost* the same value is used also by the Global Positioning System (GPS) in official computations: the official reference system of GPS is the *World Geodetic System 1984*, WGS84, which uses a reference ellipsoid that is *almost* identical to that of GRS80.

When the semi-major axis, or equatorial radius, in the GRS80 system is $a = 6378\,137.0$ m, we may readily calculate also the semi-minor axis, or polar radius, $b = a(1 - f) = 6356\,752.314$ m.

The reference ellipsoid is already a fairly accurate description of the true figure of the Earth. The deviations of mean sea level from the GRS80 reference ellipsoid are of the order of ± 100 m.

Exercises

Exercise 1 – 1: A circle on the unit sphere

   a. Let S be a circle on the surface of the unit sphere, the radius of which (as measured on the surface) is r. Calculate the length of the circumference of S.

   b. How does the above problem change, if the radius r of the disc is thought of as running along a straight line drawn through the sphere (i.e., not along the surface of the sphere)?

Exercise 1 – 2: Map projection properties

What properties are required of a map projection, if the intended use of the map is:

1. sailing on the Gulf of Finland,

2. visualizing the surface areas of states in different parts of the world,

3. visualizing distances when flying to remote holiday destinations,

4. an intelligence-gathering mission related to a multinational peacekeeping operation?

Exercise 1 – 3: Grade measurement

a. Compute, using the GRS80 reference ellipsoid parameters a and f, what is, at the North Pole and at the equator, the meridional radius of curvature M. Use Appendix C Equation C.1 on page 249.

b. Compute how long in metres is a one-degree arc, at the North Pole and at the equator.

c. Would you be able to derive in a simple way the equation in Appendix C at least for the cases of equator ($\varphi = 0°$) and pole ($\varphi = 90°$)?

Chapter 2

Popular conformal map projections

Map projections are always chosen according to their intended use. Projections may be classified by their *properties* or by their *ways of construction*. According to this classification, we have:

- cylindrical projections,

- azimuthal projections, and

- conical projections.

Cylindrical and azimuthal projections can again be divided into normal, transversal and oblique projections. See Figure 2.1.

A *cylindrical projection* is a map projection in which parallels are mapped into horizontal straight lines, and meridians into vertical straight lines. An example of a cylindrical projection is the central cylindrical projection, obtained by mapping points on the surface of the sphere along a straight line through the centre of the sphere to a cylinder touching the sphere, and then, unrolling the cylinder. The central cylindrical projection is, however, not the Mercator projection.

For general (topographic) maps, only *conformal* projections are used. Of these, the most popular are the Gauss–Krüger, a transversal Mercator projection, and the *Universal Transverse Mercator* (UTM) projection, a variant of the Gauss–Krüger projection.

If we only consider the above mentioned projection types in their normal orientations, there are three conformal basic projections:

- the conformal cylindrical projection, i.e., the Mercator projection

- the conformal azimuthal projection, the so-called stereographic projection

- the conformal conical projection or Lambert projection.

These three basic projections and the theoretical connections between them will be discussed below, in the geometry of the ellipsoid of revolution. The equations contain the following basic quantities on the ellipsoid of revolution:

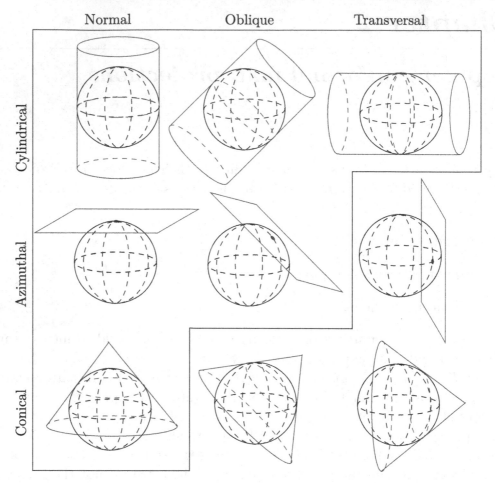

Figure 2.1: Classification of map projections. Those outside the frame are not useful in practice.

- the meridional radius of curvature $M(\varphi)$,

- the transversal radius of curvature (i.e., the radius of curvature of the ellipsoidal surface in the East–West direction) $N(\varphi)$, and

- the distance from the axis of rotation $p(\varphi) = N(\varphi)\cos(\varphi)$.

All three quantities depend only on the geodetic (or geographical) latitude φ.

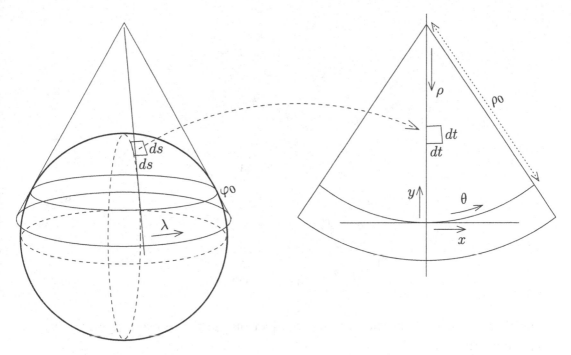

Figure 2.2: The Lambert projection.

2.1 The Lambert projection

The Lambert[1] projection (LCC, Lambert Conformal Conical) is a conformal conical projection. The scale along each parallel is constant, it is however latitude dependent, having its minimum value at a certain latitude φ_0 in the middle of the area to be mapped. Generally, this scale value is $m < 1$; then, there are *two standard parallels* on which the scale is $m = 1$. See Figures 2.2 and 2.3. The meridians are in the map plane straight lines that intersect in a single point, which also is the common centre of all parallels.

The polar co-ordinates in the map plane are ρ and θ. They are related to rectangular co-ordinates in the map plane x and y as follows:

$$x = \rho \sin \theta,$$
$$y = \rho_0 - \rho \cos \theta,$$

in which $\rho = \rho_0$ refers to the parallel $\varphi = \varphi_0$.

Also

$$\theta = n\lambda, \tag{2.1}$$

in which n must be chosen suitably.

---

[1] Johann Heinrich Lambert $(1728-1777)$ was a Swiss mathematician, physicist and astronomer.

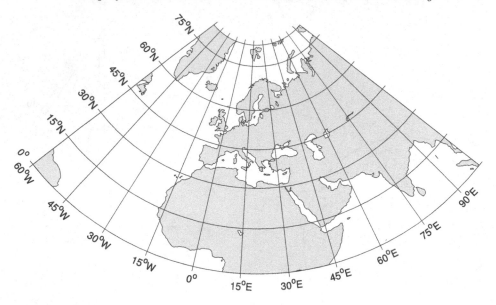

Figure 2.3: Example of the Lambert projection (software used: `m_map`, Pawlowicz maintained).

Let ds be an element of length on the Earth's surface, and dt the corresponding element in the map plane. Then, when moving along a parallel

$$ds = p(\varphi)\, d\lambda,$$
$$dt = n\rho\, d\lambda,$$

and by division one obtains

$$\frac{dt}{ds} = \frac{n\rho}{p}.$$

Based on conformality, this will also apply when moving along a meridian. Based on the definition of meridional curvature M we have $ds = M(\varphi)\, d\varphi$, i.e.,

$$\frac{dt}{d\varphi} = \frac{dt}{ds}\frac{ds}{d\varphi} = \frac{n\rho}{p}M.$$

If ρ is reckoned positive toward the South, i.e., $d\rho = -dt$, we obtain

$$\frac{d\rho}{d\varphi} = -n\rho(\varphi)\frac{M(\varphi)}{p(\varphi)}. \tag{2.2}$$

With Equations 2.1 and 2.2, we may compute (θ, ρ), if given is (φ, λ). For further treatment, we first move ρ to the left-hand side:

$$\frac{d}{d\varphi}\ln\rho = -n\frac{M}{p} \implies \rho = \rho_0 \exp\left(-n\int_{\varphi_0}^{\varphi}\frac{M(\varphi')}{p(\varphi')}d\varphi'\right).$$

If we define

$$\psi(\varphi) \stackrel{\text{def}}{=} \int_0^\varphi \frac{M(\varphi')}{p(\varphi')} d\varphi',$$

the so-called *isometric latitude*, we obtain

$$\rho = \rho_0 \exp\left(-n\big(\psi(\varphi) - \psi(\varphi_0)\big)\right) = \rho_0 \frac{\exp\big(-n\psi(\varphi)\big)}{\exp\big(-n\psi(\varphi_0)\big)}. \qquad (2.3)$$

One also should choose a value n. This is done in such a way that the *scale is stationary on the reference latitude* φ_0:

$$\frac{dm}{d\varphi} = 0 \qquad \text{if } \varphi = \varphi_0.$$

The scale is

$$m \stackrel{\text{def}}{=} \frac{d\rho}{ds} = -\frac{dt}{ds} = -\frac{n\rho}{p},$$

i.e.,

$$\frac{dm}{d\varphi} = -\frac{n}{p}\frac{d\rho}{d\varphi} + \rho\frac{n}{p^2}\frac{dp}{d\varphi} = \frac{n^2\rho}{p^2}M - \frac{n\rho}{p^2}M\sin\varphi = \frac{n\rho}{p^2}M\left(n - \sin\varphi\right), \quad (2.4)$$

using Equation 2.2 as well as the result from Appendix C on page 249:

$$\frac{dp}{d\varphi} = \frac{d}{d\varphi}\big(N(\cos\varphi)\big) = -M(\varphi)\sin\varphi.$$

The scale m is now stationary on the reference latitude $\varphi = \varphi_0$, if the expression 2.4 vanishes. That may be achieved by choosing

$$n = \sin\varphi_0.$$

As the initial condition, we still need to choose ρ_0, which give for the scale on the reference parallel

$$m(\varphi_0) = n\frac{\rho_0}{p(\varphi_0)}.$$

Alternatively one may choose the value φ_1 such, that $m(\varphi_1) = 1$ (standard latitude): then

$$n\frac{\rho_1}{p(\varphi_1)} = 1,$$

from which $\rho_1 \stackrel{\text{def}}{=} \rho(\varphi_1)$ follows. Now $\rho(\varphi)$ is obtained by integrating Equation 2.2 from the initial value, either ρ_0 or ρ_1, the interval of integration being either $\big[\varphi_0, \varphi\big]$ or $\big[\varphi_1, \varphi\big]$. For this purpose, we also have the closed formula 2.3.

The reverse operation

The situation for the inverse operation $\theta \mapsto \lambda$ is made easier by the circumstance that the inverse solution of Equation 2.2, a differential equation, i.e., computing φ if given is ρ, may be done as follows:

1. Invert analytically Equation 2.3:

$$\psi(\varphi) = -\frac{\psi(\varphi_0)}{n}\left(\ln \rho - \ln \rho_0\right);$$

2. Carry out the inverse calculation $\psi \mapsto \varphi$ (see below).

2.2 About the isometric latitude

The isometric latitude,

$$\psi = \int_0^\varphi \frac{M(\varphi')}{p(\varphi')} d\varphi',$$

can be calculated numerically (quadrature; the MATLAB®[2] QUAD routines, to be superceded by INTEGRAL). However, in the case of the sphere, an explicit expression

$$\psi = \ln \tan\left(\frac{\pi}{4} + \frac{\varphi}{2}\right), \tag{2.5}$$

exists, as may be easily proven by differentiating the expression using the chain rule.

Even on the ellipsoid of revolution, a closed solution[3] may be found:

$$\psi = \ln\left(\tan\left(\frac{\pi}{4} + \frac{\varphi}{2}\right)\left(\frac{1 - e\sin\varphi}{1 + e\sin\varphi}\right)^{e/2}\right). \tag{2.6}$$

See Appendix B on page 247.

The reverse operation

If ψ is given and one wishes to compute φ, one may use the equation

$$\frac{d\psi}{d\varphi} = \frac{M(\varphi)}{p(\varphi)}$$

---

[2]MATLAB is a trade mark of The MathWorks Inc.

[3]Note that e here denotes the first eccentricity of the Earth ellipsoid, Equation 8.4 on page 99, not the base number of the natural logarithms.

and invert it:

$$\frac{d\varphi}{d\psi} = \frac{p(\varphi)}{M(\varphi)}.$$

This equation is of general form

$$\frac{dy}{dt} = f(y, t),$$

and may be solved numerically by using the MATLAB ODE routines[4].

An alternative approach in the case of a sphere is the analytical inversion of Equation 2.5:

$$\varphi = 2\left(\arctan\exp\psi - \frac{\pi}{4}\right).$$

This is also suitable as a first iteration step in the ellipsoidal case:

$$\varphi^{(0)} = 2\left(\arctan\exp\psi - \frac{\pi}{4}\right),$$

after which

$$\varphi^{(i+1)} = 2\left(\arctan\left(\left(\frac{1 + e\sin\varphi^{(i)}}{1 - e\sin\varphi^{(i)}}\right)^{e/2}\exp\psi\right) - \frac{\pi}{4}\right).$$

This process converges rapidly.

2.3   The Mercator projection

The classical Mercator projection, Figure 2.4, is obtained as a limiting case of the Lambert projection by choosing the limit values $n \to 0$, $\rho_0 \to \infty$, but nevertheless $n\rho_0 = 1$, and also $\varphi_0 = 0$. Then

$$\rho = \rho_0\exp\left(-n\big(\psi(\varphi) - \psi(\varphi_0)\big)\right) \approx$$
$$\approx \rho_0 - n\rho_0\big(\psi(\varphi) - \psi(\varphi_0)\big) =$$
$$= \rho_0 - \psi(\varphi).$$

Choose $y \stackrel{\text{def}}{=} -(\rho - \rho_0)$ and $x \stackrel{\text{def}}{=} \lambda$ and obtain the *Mercator projection formulas*

$$x = \lambda,$$
$$y = \psi(\varphi) = \int_0^\varphi \frac{M(\varphi')}{p(\varphi')}d\varphi'. \tag{2.7}$$

---

[4]Similar functionality is found in the open-source octave software.

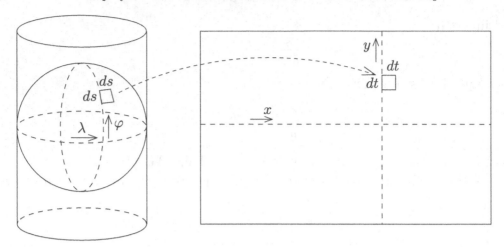

Figure 2.4: The Mercator projection.

Here, the isometric latitude shows up at its most simple: in the case of spherical geometry, we obtain

$$y = \ln \tan\left(\frac{\pi}{4} + \frac{\varphi}{2}\right).$$

Mercator is *not* a so-called lamp projection: there is no projection centre that the "light" is emanating from.

2.4 The stereographic projection

This azimuthal or plane projection is also conformal, and, like Mercator, it is also a limiting case of the Lambert projection. See Figure 2.7 for an application example.

Choose in Equation 2.3 on page 21 $n = 1$ and choose $\varphi_0 = 0$ (equator) and ρ_0 correspondingly, i.e., $\rho_0 \stackrel{\text{def}}{=} \rho(\varphi_0)$. Then

$$\rho = \rho_0 \exp\left(-\psi(\varphi)\right).$$

Unfortunately, for the limiting value $\varphi \to \frac{\pi}{2}$ the function $\psi(\varphi)$ diverges; we may define

$$\rho = \begin{cases} \rho_0 \exp\left(-\psi(\varphi)\right) & \text{if } \varphi < \frac{\pi}{2}, \\ 0 & \text{if } \varphi = \frac{\pi}{2}, \end{cases}$$

after which $\rho(\varphi)$ is continuous.

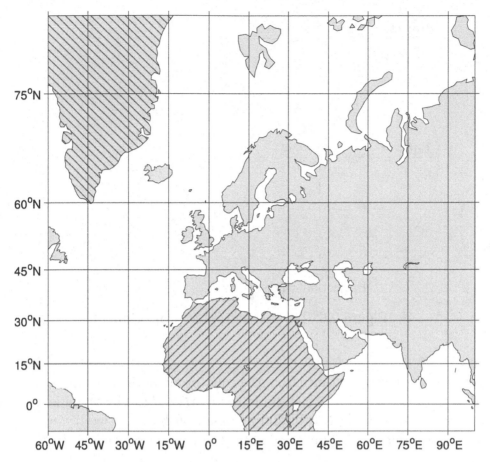

Figure 2.5: Example of the Mercator projection. Note that Greenland and Africa look approximately of the same size, although the area of Greenland, 2.2 million square kilometres, is only 7% of Africa's 30 million square kilometres.

On the sphere we obtain

$$\rho = \frac{\rho_0}{\tan\left(\frac{\pi}{4} + \frac{\varphi}{2}\right)} = \rho_0 \cot\left(\frac{\pi}{4} + \frac{\varphi}{2}\right) =$$

$$= \rho_0 \tan\left(\frac{\pi}{4} - \frac{\varphi}{2}\right) = \rho_0 \tan\left(\frac{1}{2}\left(\frac{\pi}{2} - \varphi\right)\right).$$

This case has been visualized in Figure 2.6. The number ρ_0 is the distance from the centre of the projection ("South Pole") to the projection plane. In this case, the projection is a real "lamp projection"... but unfortunately *only* for spherical geometry. Let us derive the *scale* of the stereographic projection:

$$\frac{d\rho}{ds} = \frac{d\rho}{d\psi}\frac{d\psi}{d\varphi}\frac{d\varphi}{ds} = -\rho_0 \exp(-\psi) \cdot \frac{M}{p} \cdot \frac{1}{M} = -\frac{\rho}{p} = -\rho_0 \frac{\exp\left(-\psi(\varphi)\right)}{p(\varphi)},$$

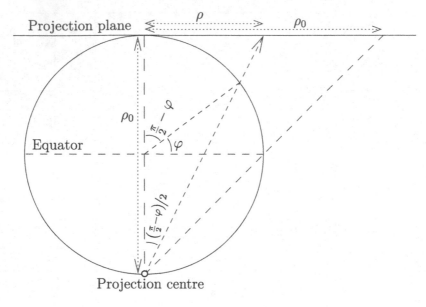

Figure 2.6: The stereographic projection.

in which s is the metric distance from the equator along the meridian, measured on the surface of the reference ellipsoid, $ds = M(\varphi)d\varphi$. The value is negative, because s increases going North, and ρ going South.

Next, use Equation 2.6 on page 22 for the isometric latitude. Recall that

$$e^{-\psi} = \left(e^{\psi}\right)^{-1} = \left(\tan\left(\frac{\pi}{4} + \frac{\varphi}{2}\right)\left(\frac{1 - e\sin\varphi}{1 + e\sin\varphi}\right)^{e/2}\right)^{-1} =$$

$$= \tan\left(\frac{\pi}{4} - \frac{\varphi}{2}\right)\left(\frac{1 - e\sin\varphi}{1 + e\sin\varphi}\right)^{-e/2}.$$

From this follows

$$\frac{d\rho}{dx} = -\frac{\rho_0}{N\cos(\varphi)}\tan\left(\frac{\pi}{4} - \frac{\varphi}{2}\right)\left(\frac{1 - e\sin\varphi}{1 + e\sin\varphi}\right)^{-e/2}.$$

Close to the pole, we have

$$\tan\left(\frac{\pi}{4} - \frac{\varphi}{2}\right) \approx \frac{1}{2}\cdot\left(\frac{\pi}{2} - \varphi\right)$$

and

$$\cos\varphi = \sin\left(\frac{\pi}{2} - \varphi\right) \approx \frac{\pi}{2} - \varphi$$

as well as

$$N = \frac{a}{\sqrt{1 - e^2}}.$$

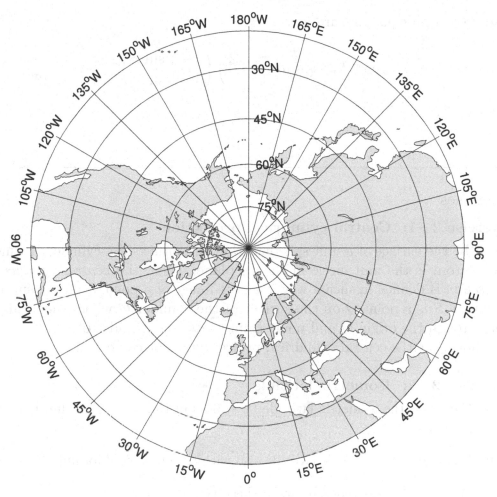

Figure 2.7: Example of stereographic projection.

We obtain as the end result close to the pole

$$\frac{d\rho}{dx} \approx -\frac{\rho_0}{2a}\sqrt{1-e^2}\left(\frac{1-e}{1+e}\right)^{-e/2}.$$

For the GRS80 reference ellipsoid, we have $e = 0.081\,819\,191\,042\,810\,976\,93$. With this, we obtain for the expression in e on the right-hand side

$$\sqrt{1-e^2}\left(\frac{1-e}{1+e}\right)^{-e/2} = 1.003\,356\,555\,2658.$$

If we want the scale to be 1 at the pole, we must set

$$\rho_0 = 1.993\,309\,347\,0149a.$$

The projection equations are now

$$\rho = \rho_0 e^{-\psi} = \rho_0 \tan\left(\frac{\pi}{4} - \frac{\varphi}{2}\right)\left(\frac{1 - e\sin\varphi}{1 + e\sin\varphi}\right)^{-e/2},$$

$$\theta = \lambda.$$

Exercises

Exercise 2 – 1: Central cylindrical projection

Let S be the surface of a unit sphere, and C the surface of a cylinder of which the bottom is the unit circle in the xy plane. Then, the *central cylindrical projection* is defined in such a way, that the image of a point P on the surface of sphere S is a point Q on the surface of the cylinder C where the straight line through the origin O and point P intersects the cylinder C.

Show that the central cylindrical projection is not conformal.

Exercise 2 – 2: Isometric latitude

a. Calculate the isometric latitude ($\varphi = 60°$) of Helsinki using the spherical formula.

b. Calculate the isometric latitude also with the ellipsoidal formula (GRS80).

c. What is the difference between these two?

Chapter 3

The complex plane and conformal mappings

3.1 Complex numbers

The term *imaginary number* was used for the first time by the French philosopher and mathematician René Descartes in his work *La Géométrie* (1637). Descartes intended the term to be derogatory, as he did not believe in the existence of imaginary numbers[1]. This was caused by Descartes thinking of complex numbers first and foremost as solutions of "unsolvable" polynomial equations like $x^2 + 1 = 0$. *La Géométrie* was originally a supplementary appendix to the philosophical work *Discours de la Méthode* ("Discourse on Method"). It has nevertheless had much influence on mathematics, e.g., on differential and integral calculus.

Figure 3.1: Leonhard Euler, painting by Johann Georg Brucker, around 1756.

The Swiss mathematician Leonhard Euler (1707–1783) in the 18th century (see Figure 3.1) took into use the symbol i for the imaginary unit, but it was the geometrical interpretation of complex numbers invented by the Norwegian Caspar Wessel[2] and popularized by Carl Friedrich Gauss which

---

[1]In the history of mathematics, the modern view that mathematics is not so much a branch of natural science as a formal language serving the needs of inference and computation, has not always been held. According to this view, a certain type of number exists once it has been defined: the natural numbers $\mathbb{N}$ are the number zero plus, for every natural number n, its successor number $n + 1$; integer and rational numbers are equivalence classes of natural number pairs; etc.

[2]Caspar Wessel (1745–1818) was a Norwegian-Danish mathematician, cartographer and land surveyor.

led to their general acceptance. Gauss also introduced the name "complex number."[3]

A complex number may be presented in the form $z = x + iy$, where the imaginary unit i satisfies the equation $i^2 = -1$ and x, y are real numbers. The number $\Re\, z = x$ is the real part of complex number z and $\Im\, z = y$ is its imaginary part. The complex numbers $z = a + ib$ and $w = c + id$ are identical if and only if both $a = c$ and $b = d$. In particular, the complex number $z = a + ib$ is zero if and only if both $a = 0$ and $b = 0$.

Unlike for real numbers, the comparison operators $<, \leq$ are not defined for complex numbers. A set of complex numbers can thus not be arranged into ascending order.

Let $z = a + ib$ and $w = c + id$ be complex numbers. Then we obtain arithmetic operations as follows.

- Sum: $$z + w = (a + c) + i(b + d).$$

- Difference: $$z - w = (a - c) + i(b - d).$$

- Product: $$zw = (a + ib)(c + id) = (ac - bd) + i(ad + bc).$$

- Quotient: $$\frac{z}{w} = \frac{a + ib}{c + id} \cdot \frac{c - id}{c - id} = \left(\frac{ac + bd}{c^2 + d^2}\right) + i\left(\frac{bc - ad}{c^2 + d^2}\right).$$

The addition and subtraction of complex numbers correspond to operations on vectors. From these computation rules, we obtain immediately the following formulas for the powers of the imaginary unit:

$$i^2 = -1, \qquad i^3 = -i, \qquad i^4 = 1, \qquad \dots,$$

$$\frac{1}{i} = -i, \qquad \frac{1}{i^2} = -1, \qquad \frac{1}{i^3} = i, \qquad \dots.$$

If $z = a + 0i$, then z is a real number. All formulas given above are also true for real numbers. So, if the imaginary part is zero, the formulas revert to the familiar properties of real numbers.

The *complex conjugate* of the complex number $z = x + iy$, $\bar{z}$, is defined by the formula $\bar{z} = x - iy$. The *modulus* or *norm* of the number is

$$r \overset{\text{def}}{=} |z| = \sqrt{x^2 + y^2} = \sqrt{z\bar{z}}.$$

For multiplication and division, the following holds:

$$|zw| = |z||w|, \qquad \left|\frac{z}{w}\right| = \frac{|z|}{|w|}.$$

---

[3] Actually Gauss did not like the word "imaginary" with its negative connotation, and proposed even the name "sideways quantities" (Gauss, 1876).

3.2 The polar representation of complex numbers

A complex number $z = x + iy$ can be identified with the point (x, y) in the plane, which is also called the *complex plane* (see Figure 3.2). The *argument* or *phase angle* $\arg z$ of a complex number $z = x + iy$, $z \neq 0$ corresponds to the angle, at the origin, between the point corresponding to the number z and the positive half of the x-axis, which is also called the *real axis*. The argument is defined via standard trigonometry by

$$\tan(\arg z) \overset{\text{def}}{=} \frac{y}{x}, \qquad x \neq 0.$$

We note that $\arg z$ is *multivalued* because $\tan$ is a periodic function with period π and thus, like all trigonometric functions, with period 2π. Observe that, for example, $z = x + iy$ and $-z = -x - iy$ have different arguments, despite the fact that the ratio in the above formula is the same. For this reason, when computing the argument, it is necessary to consider the quadrant of the complex plane where the point lies. In addition, the case $x = 0$ must be treated separately.

If $y > 0$, we define $\arg z = \frac{\pi}{2} + 2n\pi$, where n is any integer. Similarly if $y < 0$, we set $\arg z = -\frac{\pi}{2} + 2n\pi$. The argument of the complex number 0 remains undefined.

Because the argument of a complex number is not unique, we define the *principal value* of the argument is called that value of the argument of a complex number, that lies in the interval

$$-\pi < \theta \overset{\text{def}}{=} \operatorname{Arg} z \leq +\pi.$$

Generally it holds that

$$\arg z = \theta + 2n\pi,$$

where θ is the principal value and n is any integer.

For complex numbers, it also holds (*triangle inequality*) that

$$|z + w| \leq |z| + |w|.$$

The triangle inequality can be proven by the following reasoning. Compute

$$|z + w|^2 = (z + w)\overline{(z + w)}$$
$$= z\bar{z} + z\bar{w} + \bar{z}w + w\bar{w}$$
$$= |z|^2 + z\bar{w} + \bar{z}w + |w|^2.$$

Because $\overline{z\bar{w}} = \bar{z}\bar{\bar{w}} = \bar{z}w$, and thus

$$z\bar{w} + \bar{z}w = z\bar{w} + \overline{z\bar{w}} = 2\,\mathfrak{Re}(z\bar{w}) \leq 2|z\bar{w}| = 2|z||\bar{w}| = 2|z||w|,$$

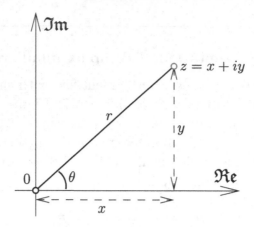

Figure 3.2: The complex plane.

we obtain
$$|z + w|^2 \leq |z|^2 + 2|z||w| + |w|^2 = (|z| + |w|)^2.$$

From Figure 3.2 we see that
$$x = r \cos \theta,$$
$$y = r \sin \theta.$$

So
$$z = x + iy = r \cos \theta + ir \sin \theta.$$

We obtain the representation of the complex number *in polar form*:
$$z = r(\cos \theta + i \sin \theta).$$

For the product formula of the complex numbers
$$z = r \left(\cos \theta + i \sin \theta \right),$$
$$w = s \left(\cos \phi + i \sin \phi \right),$$

we obtain
$$zw = rs \left[(\cos \theta \cos \phi - \sin \theta \sin \phi) + i \left(\cos \theta \sin \phi + \sin \theta \cos \phi \right) \right],$$

i.e., with the familiar trigonometric formulas
$$zw = rs \Big(\cos(\theta + \phi) + i \sin(\theta + \phi) \Big).$$

In multiplying complex numbers, the moduli are multiplied and the arguments added together.

Here we already begin to see what makes complex numbers so useful: they offer a natural way of *describing rotations as multiplications*. This is to a large extent what the extensive use of complex numbers in physics and engineering is based on, always when we are describing rotational motions — or oscillatory phenomena like electromagnetism. An oscillation is one component of a rotational motion.

3.3   On the nature of the exponential function

Often, the definition of the base of the exponential function e is given as

$$e = \lim_{n \to \infty} \left(1 + \frac{1}{n}\right)^n.$$

For Napier's constant e, we obtain the following approximations:

- $n = 100$: $e \approx 1.01^{100} = 2.704\,8138$,

- $n = 1000$: $e \approx 1.001^{1000} = 2.716\,9239$,

- $n = 10\,000$: $e \approx 1.0001^{10,000} = 2.718\,1459$,

when the exact value is $e = 2.718\,2818\ldots$. We may say that e is obtained by multiplying with each other very many numbers that differ only little (in fact, infinitesimally little) from the number one. The scaling is repeated while the scale factor is close to unity.

We may write more generally:

$$e^x = \left[\lim_{n \to \infty} \left(1 + \frac{1}{n}\right)^n\right]^x = \lim_{n \to \infty} \left(1 + \frac{1}{n}\right)^{nx} = \lim_{n \to \infty} \left(1 + \frac{x}{n}\right)^n.$$

The latter identity is probably not obvious. At least for rational numbers $x \in \mathbb{Q}$ it can be easily proven: let $x = \frac{p}{q}, p \in \mathbb{Z}, q \in \mathbb{N}$. Then, for the sub-sequences of the sequence $n = 1, 2, 3, \ldots$ given by $n = q, 2q, 3q, \ldots xn$ (or, if $x < 0$, $-xn$) is a natural number, let us say k. And if the expression converges for the sequence of indices n, then it will also do so for the sequence of indices k, and to the same limit value:

$$\lim_{n \to \infty} \left(1 + \frac{1}{n}\right)^{nx} = \lim_{k \to \infty} \left(1 + \frac{x}{k}\right)^k,$$

from which

$$\lim_{k \to \infty} \left(1 + \frac{x}{k}\right)^k = \left[\lim_{n \to \infty} \left(1 + \frac{1}{n}\right)^n\right]^x = e^x.$$

For irrational numbers $x \in \mathbb{R} \setminus \mathbb{Q}$, the proof is difficult[4].

In connection with complex numbers, we study the following limit:

$$e^i = \lim_{n \to \infty} \left(1 + \frac{i}{n}\right)^n.$$

Numerically this is computed thusly:

- $n = 100$: $e^i \approx (1 + 0.01i)^{100} = 0.543\,04 + 0.845\,67i$,

- $n = 1000$: $e^i \approx (1 + 0.001i)^{1000} = 0.540\,57 + 0.841\,89i$,

- $n = 10\,000$: $e^i \approx (1 + 0.0001i)^{10,000} = 0.540\,33 + 0.841\,51i$.

From this we may compute modulus and argument: for the case $n = 10\,000$ we obtain

$$\left|e^i\right| = \sqrt{0.540\,33^2 + 0.841\,51^2} = 1.0000,$$

$$\operatorname{Arg} e^i = \arctan \frac{0.841\,51}{0.540\,33} = 1.0000.$$

From this follows that e^i is a pure rotational motion, to the value of one radian. Based on the formula $e^i = (e^{i/m})^m$, $m = 1, 2, \ldots$, this rotation may be represented by using very small rotations.

Based on this we see also, that e^{it}, $t \in \mathbb{R}$, represents a rotation by an angle t: $\arg e^{it} = t$ (if the correct branch is chosen). It holds generally for complex numbers $z \neq 0$ that

$$z = e^{u+it} = e^u e^{it},$$

where e^u is the modulus of the number z and t its argument.

---

[4]In the proof, the axiomatics of real numbers is required: the set of real numbers can be uniquely sorted into an ascending order, and if a non-empty set is bounded from above, then there exists a real number called the *smallest upper bound* or *supremum* (Dedekind completeness). Let there be given a rational valued, strictly ascending sequence $q_m, m \in \mathbb{N}$, which is bounded from above. It has a smallest upper bound x, which is at the same time the limit: $x = \lim_{m \to \infty} q_m$. Then also the sequences

$$\lim_{n \to \infty} \left(1 + \frac{1}{n}\right)^{n q_m} \quad \text{and} \quad \lim_{n \to \infty} \left(1 + \frac{q_m}{n}\right)^n, \quad m \in \mathbb{N}$$

are strictly ascending, and *identical*. So, the smallest upper bounds of both, the limits

$$\lim_{m \to \infty} \lim_{n \to \infty} \left(1 + \frac{1}{n}\right)^{n q_m} = \lim_{n \to \infty} \left(1 + \frac{1}{n}\right)^{n x}$$

$$\text{and} \quad \lim_{m \to \infty} \lim_{n \to \infty} \left(1 + \frac{q_m}{n}\right)^n = \lim_{n \to \infty} \left(1 + \frac{x}{n}\right)^n,$$

are also identical.

3.4   Euler's formula

For the exponential function and for the trigonometric functions, the following series expansions are valid for real numbers $x \in \mathbb{R}$:

$$e^x = 1 + x + \frac{x^2}{2!} + \frac{x^3}{3!} + \ldots + \frac{x^n}{n!} + \ldots$$

$$\sin x = x - \frac{x^3}{3!} + \frac{x^5}{5!} - \ldots + \frac{(-1)^k x^{2k+1}}{(2k+1)!} + \ldots$$

$$\cos x = 1 - \frac{x^2}{2!} + \frac{x^4}{4!} - \ldots + \frac{(-1)^k x^{2k}}{(2k)!} + \ldots$$

If we assume at least the e^x expansion to also apply for complex numbers, it follows from these series expansions that

$$\begin{aligned}
e^{ix} &= 1 + ix + \frac{(ix)^2}{2!} + \frac{(ix)^3}{3!} + \ldots \\
&= 1 + ix + \frac{i^2 x^2}{2!} + \frac{i^3 x^3}{3!} + \ldots \\
&= 1 - \frac{x^2}{2!} + \frac{x^4}{4!} + \ldots + i\left(x - \frac{x^3}{3!} + \frac{x^5}{5!} - \ldots\right) \\
&= \cos x + i \sin x.
\end{aligned}$$

We obtain *Euler's formula*:

$$e^{i\theta} = \cos \theta + i \sin \theta. \tag{3.1}$$

As a result of Euler's formula 3.1, we obtain the following identities for trigonometric functions. Because

$$\begin{aligned}
e^{-i\theta} &= \cos(-\theta) + i \sin(-\theta) \\
&= \cos(\theta) - i \sin(\theta),
\end{aligned}$$

it holds that

$$\cos \theta = \frac{e^{i\theta} + e^{-i\theta}}{2}, \qquad \sin \theta = \frac{e^{i\theta} - e^{-i\theta}}{2i}.$$

Generally we may write for a complex number $z = x + iy$

$$\cos z = \frac{1}{2}(e^{iz} + e^{-iz}), \qquad \sin z = \frac{1}{2i}(e^{iz} - e^{-iz}).$$

Furthermore, we may also define

$$\tan z = \frac{\sin z}{\cos z}, \qquad \cot z = \frac{\cos z}{\sin z}.$$

Almost all properties of trigonometric functions familiar from the real domain apply also for complex numbers. The *hyperbolic functions* are defined as follows:

$$\cos(iz) = \cosh z, \qquad \sin(iz) = i \sinh z.$$

We obtain the formulas

$$\cosh z = \frac{e^z + e^{-z}}{2}, \qquad \sinh z = \frac{e^z - e^{-z}}{2}.$$

As earlier, we may also define

$$\tanh z = \frac{\sinh z}{\cosh z}, \qquad \coth z = \frac{\cosh z}{\sinh z}.$$

An important application of Euler's formula is the geometric interpretation of complex multiplication. Apply Euler's formula to multiplication of complex numbers:

$$w = z_1 z_2 = r_1 e^{i\theta_1} \cdot r_2 e^{i\theta_2} = (r_1 r_2) e^{i(\theta_1 + \theta_2)}.$$

So, in multiplying complex numbers, as already shown earlier in a different way, the moduli are multiplied $|z_1 z_2| = |z_1||z_2|$ and the arguments are added $\arg(z_1 z_2) = \arg(z_1) + \arg(z_2)$.

As an application of the above formulas, we also obtain useful identities for the exponential function

$$|e^{i\theta}| = |\cos\theta + i\sin\theta| = \sqrt{\cos^2\theta + \sin^2\theta} = 1.$$

So

$$|e^{i\theta}| = 1.$$

Because $e^z = e^{x+iy} = e^x e^{iy} = e^x(\cos y + i\sin y)$, we obtain

$$|e^z| = e^x, \qquad\qquad \arg(e^z) = y, \qquad\qquad e^{i\,2\pi} = 1,$$
$$e^{i\frac{\pi}{2}} = i, \qquad\qquad e^{i\pi} = -1, \qquad\qquad e^{-i\frac{\pi}{2}} = -i,$$
$$e^{z+i\,2\pi} = e^z e^{i\,2\pi} = e^z.$$

3.5   De Moivre's formula

Derive an expression for the integer powers of a complex number:

$$z^n = (re^{i\theta})^n = r^n e^{i\,(n\theta)} = r^n(\cos n\theta + \sin n\theta).$$

In particular, if $r = 1$, we obtain the result known as De Moivre[5]'s formula:

$$(\cos\theta + i\sin\theta)^n = \cos n\theta + i\sin n\theta. \tag{3.2}$$

---

[5] Abraham de Moivre (1667–1754) was a French-born mathematician and Huguenot who moved to England.

De Moivre's formula 3.2 is especially useful when seeking the nth roots of the complex number $z_0 \neq 0$. If $z^n = z_0$, we may write $z = re^{i\theta}$ and $z_0 = r_0 e^{i\theta_0}$, and we obtain

$$r^n e^{in\theta} = r_0 e^{i\theta_0},$$

i.e.,

$$r = \sqrt[n]{r_0} \quad \text{and} \quad n\theta = \theta_0 + 2k\pi,$$

in which $r = \sqrt[n]{r_0}$ is the nth root of the positive real number r_0.

All the nth roots of the number z are thus obtained from the expression

$$\sqrt[n]{|z_0|} \exp \frac{i(\theta_0 + 2k\pi)}{n}, \tag{3.3}$$

in which k is any integer. We see also that every $k = 0, 1, \ldots, n-1$ gives a different value, but the other values of k only repeat some of the earlier ones, because $e^{2\pi i k} = 1$. So, the complex number $z_0 \neq 0$ has precisely n different n^{th} roots.

From Equation 3.3 we see also that all roots have the same modulus $\sqrt[n]{|z_0|}$, and the arguments are evenly spaced. Therefore, all roots are located on a circle centred on the origin, the radius of which is $\sqrt[n]{|z_0|}$. Thus, if $z = re^{i\theta} \neq 0$, the equation $w^n = z$ has precisely n different solutions, which are obtained from the equation

$$w_k = \sqrt[n]{r} \exp \frac{i(\theta + 2k\pi)}{n},$$

in which $k = 0, 1, \ldots, n-1$, $\sqrt[n]{r}$ is the positive n^{th} root of the number $r = |z|$, and $\theta = \text{Arg } z$.

In particular, the nth roots of the number 1 are obtained from the equation

$$\omega_k = \exp \frac{i2k\pi}{n}, \qquad k = 0, 1, \ldots, n-1.$$

Example. Figure 3.3 shows the fifth roots of unity.

3.6   The logarithm function

It is natural to think of the logarithm function as the inverse mapping of the exponential function. Recall the following basic properties of the exponential function for real numbers:

- $e^x > 0$ for all $x \in \mathbb{R}$,

- $e^x \to \infty$, when $x \to \infty$,

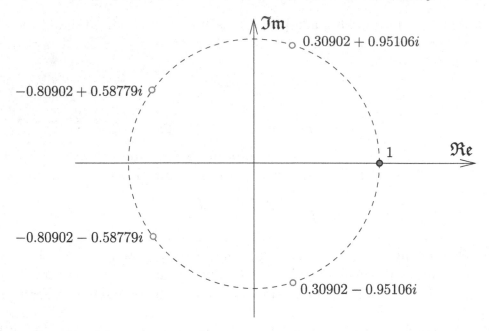

Figure 3.3: The fifth roots of unity, i.e., all different solutions of the equation $z^5 - 1 = 0$.

○ $e^{-x} = \dfrac{1}{e^x}$ (so $e^x \to 0$, when $x \to -\infty$),

○ $\dfrac{\partial}{\partial x} e^x = e^x$, and thus e^x is strictly ascending.

The exponential function e^x is thus a continuous, strictly ascending, and differentiable function from $\mathbb{R}$ to the set $\mathbb{R}_+ \overset{\text{def}}{=} \{x \in \mathbb{R} : x > 0\}$. Thus it has a continuous and strictly ascending inverse mapping, the (natural) logarithm (base number e)

$$\ln\colon \mathbb{R}_+ \to \mathbb{R},$$

for which holds $\ln x = y$ and which is the solution of the equation $e^y = x$. In particular, for each $x > 0$ there exists precisely one such y, for which $e^y = x$.

In the same way, we may define the logarithm of a complex number z as the solution $w \in \mathbb{C}$ of the equation $e^w = z$, i.e., we write $w = \ln z$, if $e^w = z$. Because $e^w \neq 0$ for all $w \in \mathbb{C}$, the number 0 has no logarithm.

Consider an arbitrary complex number $z \neq 0$ in polar form

$$z = |z| e^{\operatorname{Arg} z} = r e^{i\theta} \qquad (r = |z| > 0,\ -\pi < \theta \leq \pi).$$

Solve the equation $w = \ln z$. If we write $w = x + iy\ (x, y \in \mathbb{R})$, then we get the equation $e^w = z$ into the form $e^{x+iy} = r e^{i\theta}$, so that

$$e^x = r,\ e^{(y-\theta)i} = 1,\ \text{i.e.,}\ x = \ln r,\ y = \theta + 2k\pi, \qquad k \in \mathbb{Z}.$$

We thus obtain the following equation for the logarithm of the complex number: for $z \neq 0$,

$$w = \ln z = \ln |z| + i(\operatorname{Arg} z + 2k\pi), \qquad k \in \mathbb{Z}.$$

The *principal value of the logarithm* (notation Ln) corresponds to the principal value of the argument of a complex number. In other words, if $z \neq 0$, then

$$\operatorname{Ln} z = \ln |z| + i \operatorname{Arg} z, \qquad -\pi < \operatorname{Arg} z \leq \pi.$$

If z is a positive real number (so $\operatorname{Arg} z = 0$), this corresponds to the familiar meaning of the notation "$\ln z$."

Examples.

- $\operatorname{Ln}(\pm i) = \pm i \cdot \frac{\pi}{2}$,

- $\operatorname{Ln}(1 + i) = \ln \sqrt{2} + i \cdot \frac{\pi}{4}$,

- $\operatorname{Ln} \frac{1 \pm i}{\sqrt{2}} = \pm i \cdot \frac{\pi}{4}$,

- $\operatorname{Ln}(-1) = i\pi$,

- $\operatorname{Ln}(i^{1/4}) = i \cdot \frac{\pi}{8}$,

- $\operatorname{Ln}(\alpha z) = \ln \alpha + \operatorname{Ln} z, \ (\alpha > 0)$.

Generally for the complex numbers z_1, z_2, unlike in the case of real numbers, it is not true that

$$\operatorname{Ln}(z_1 z_2) = \operatorname{Ln} z_1 + \operatorname{Ln} z_2,$$

even though we would only be looking at the principal value of the logarithm. If $z_1, z_2 \neq 0$, it holds nevertheless that

$$\ln(z_1 z_2) = \ln z_1 + \ln z_2 \quad \text{"modulo } 2\pi i\text{,"}$$

i.e.,

$$\ln(z_1 z_2) = \ln z_1 + \ln z_2 + 2k\pi i,$$

in which $k \in \mathbb{Z}$ is an unknown (but usually small) integer. We use the following notation:

$$\ln(z_1 z_2) = \ln z_1 + \ln z_2 \quad (\text{mod } 2\pi i).$$

Similarly also

$$\ln\left(\frac{z_1}{z_2}\right) = \ln z_1 - \ln z_2 \quad (\text{mod } 2\pi i).$$

Exercises

Exercise 3 − 1:

Let $z_1 = 6 - 3i$, $z_2 = 3 + 6i$. Calculate z_1/z_2.

Exercise 3 – 2:

Let $z_1 = 6 - 3i$, $z_2 = 3 + 6i$. Calculate $(z_1 - \bar{z}_2)^2$.

Exercise 3 – 3:

Determine $\mathfrak{Re}(z/\bar{z})$, when $z = x + iy \neq 0$ $(x, y \in \mathbb{R})$. Give the answer as an expression in x and y.

Exercise 3 – 4:

Give the complex number $\sqrt{2}\, e^{i\pi/4}$ in the form $x + iy$.

Exercise 3 – 5:

Draw a graph of the sets

a. $\{z : \mathfrak{Im}\, z > 1\}$,

b. $\{z : |\operatorname{Arg} z| \leq \pi/4\}$,

c. $\{z : r \leq |z - a| < s\}$.

Are the sets open, closed or neither?

Exercise 3 – 6:

Express in polar form

a. -7,

b. $\dfrac{2 + 2i}{1 - i}$,

c. $\dfrac{i\sqrt{2}}{3 + 3i}$.

Exercise 3 – 7:

Find all solutions of the following equations:

a. $z^2 + z + 1 = i$,

b. $z^2 - 6z - 3 = i$.

Exercise 3 – 8:

Express the formula of De Moivre using $\cos 3\varphi$ and $\sin 3\varphi$ with the aid of powers of cos and sin. Here we assume that $\varphi \in \mathbb{R}$.

Exercise 3 – 9:

Calculate $\operatorname{Ln} z$, when z is:

   a. -10,

   b. $1 - i$,

   c. $-ei$.

Exercise 3 – 10:

Prove the following identities, when $z_1, z_2 \in \mathbb{C}$:

   a. $|1 + z_1 \bar{z}_2|^2 + |z_1 - z_2|^2 = \left(1 + |z_1|^2\right)\left(1 + |z_2|^2\right)$,

   b. $|1 - z_1 \bar{z}_2|^2 - |z_1 - z_2|^2 = \left(1 - |z_1|^2\right)\left(1 - |z_2|^2\right)$.

Exercise 3 – 11:

Find those points z in the plane, for which $\frac{1}{i-5z}$ is real-valued.

Exercise 3 – 12:

   a. Calculate $(1 + i)^{10}$,

   b. Express $\cos^4 \theta$ with the help of the following functions:

$$\cos n\theta, n = 0, 1, \ldots, 4.$$

Exercise 3 – 13:

Describe the following subsets of the plane:

   a. $\{z : \operatorname{Im} z \in (1, 2)\}$,

   b. $\{z : \operatorname{Re} z \in (-1, 1)\}$,

   c. $\{z : \operatorname{Arg} z \in (1, 2)\}$,

   d. $\{z : \operatorname{Arg}(z - 1 - 2i) \in (0, 2)\}$,

   e. $\{re^{i\varphi} : r = (1 + \cos \varphi), \varphi \in (0, 2\pi)\}$.

Exercise 3 – 14:

How does the mapping

   a. $z \mapsto 1/z$,

   b. $z \mapsto 1/\bar{z}$

map the part of the plane above the straight line $y = \dfrac{x}{2} + 1$? And how is the circle $\left|z - \dfrac{i}{2}\right| = \dfrac{1}{2}$ mapped?

Exercise 3 – 15:

Find the largest and smallest values of the expression $|z^2 - iz + 1|$ on the circle $|z| = 1$.

Exercise 3 – 16:

Express the function $f(z) = z^3$ into the form $f(z) = u(x, y) + iv(x, y)$.

Exercise 3 – 17:

Express the function $f(z) = 1/z^2$ into the form $f(z) = u(x, y) + iv(x, y)$.

Exercise 3 – 18:

Calculate the complex derivative of the function $f(z) = \dfrac{1 + z}{z - i}$.

Exercise 3 – 19:

Write the complex number $\cosh(-6 - 3i)$ into the form $x + iy$.

Chapter 4

Complex analysis

The usefulness of complex numbers in research in mathematical geodesy should be clear based on what we saw above. One central result is that any degree n polynomial of one variable with complex coefficients will have n roots (which may be multiple). The situation with respect to real roots is a different one: already a quadratic equation has either two roots, one (double) root or no roots at all.

An extremely useful application of complex numbers is obtained when they are combined with integral and differential calculus. This leads to a mathematical theory called *complex analysis*.

4.1 Limit and continuity of a complex function

We say that a sequence of complex numbers $z_1, z_2, \dots$ has a *limit* z, and we write

$$\lim_{n \to \infty} z_n = z,$$

if, for all $\varepsilon > 0$, there exists such an $N \in \mathbb{N}$, that

$$|z_n - z| < \varepsilon, \text{ when } n > N.$$

See Figure 4.1. In particular, if (z_n) is a sequence of complex numbers, then $\lim_{n \to \infty} z_n = z$, if and only if $\lim_{n \to \infty} \mathfrak{Re}\, z_n = \mathfrak{Re}\, z$ and $\lim_{n \to \infty} \mathfrak{Im}\, z_n = \mathfrak{Im}\, z$.

Introduce the following notations:

$$B(z, r) = \{w : |z - w| < r\} \qquad \text{(open disk)},$$
$$\overline{B}(z, r) = \{w : |z - w| \le r\} \qquad \text{(closed disk)}.$$

We say that the set $D \subset \mathbb{C}$ is *open*, if for every $z \in D$ there exists such an $r > 0$, that $B(z, r) \subset D$. The set $E \subset \mathbb{C}$ is *closed*, if its complement $D = \mathbb{C} \setminus E$ is open. The set $D \subset \mathbb{C}$ is an (open) *neighbourhood* of point $z \in \mathbb{C}$, if D is open and $z \in D$. The set $D \subset \mathbb{C}$ is *connected*, if two points $z, w \in D$ can always be connected by a polygonal chain in the set D. The set $D \subset \mathbb{C}$ is a *domain*, if it is open and connected.

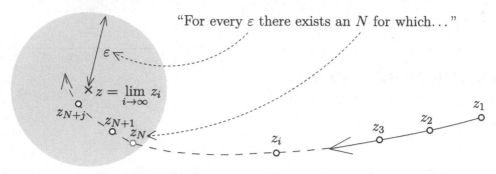

Figure 4.1: The definition of the limit, illustrated in the complex plane.

Let S be a set of complex numbers and f a complex-valued function defined on set S. Then we use the notation

$$w = f(z) = u(x, y) + iv(x, y),$$

where $u: S \to \mathbb{R}$ s the real part of function f and $v: S \to \mathbb{R}$ its imaginary part. So, a complex function is a pair of real-valued functions u, v, which both depend on two real parameters x, y, and $z = x + iy$.

Example. Let $w = f(z) = z^2 + 3z$. Find the functions u, v and compute $f(1 + 3i)$. Write $z = x + iy$, obtaining

$$f(z) = f(x + iy) = (x + iy)^2 + 3(x + iy) = x^2 + 3x - y^2 + i(2xy + 3y).$$

Thus

$$u(x, y) = x^2 + 3x - y^2 \text{ and } v(x, y) = 2xy + 3y.$$

We also obtain

$$f(1 + 3i) = (1 + 3i)^2 + 3(1 + 3i) = 1 + 6i - 9 + 3 + 9i = -5 + 15i.$$

The function $f: D \to \mathbb{C}$ has the limit c in the point z_0 — write this as $\lim\limits_{z \to z_0} f(z) = c$ — if, for all $\varepsilon > 0$, there exists such a $\delta > 0$, that, if $|z - z_0| < \delta$ and $z \in D$, then $|f(z) - c| < \varepsilon$. The function f is *continuous* in point z_0, if f is defined in some neighbourhood of point z_0, and $f(z_0) = \lim\limits_{z \to z_0} f(z)$.

For example, polynomials, the exponential function, the sine and the cosine are continuous throughout the complex plane. On the other hand, the function

$$f(z) = \begin{cases} \dfrac{\mathfrak{Re}(z^2)}{|z|^2}, & z \neq 0 \\ 0, & z = 0 \end{cases}$$

is not continuous in point $z = 0$: if one approaches the origin from the direction of the real axis, the limit obtained is 1. On the other hand, if one approaches from the imaginary axis, one obtains the limit -1. For other directions of approach, similarly other limit values are obtained. Also the function $f(z) = \text{Arg } z$, $z \neq 0$, $f(0) = 0$ is not continuous in point $z = 0$.

4.2 The complex derivative and analytic functions

We know that the derivative of a real function

$$\lim_{h \to 0} \frac{f(x_0 + h) - f(x_0)}{h}$$

does not always exist. A simple example of this is the continuous function $|x|$, for which holds, for the argument value $x_0 = 0$:

$$\lim_{h \downarrow 0} \frac{|x_0 + h| - |x_0|}{h} = 1, \quad \text{but} \quad \lim_{h \uparrow 0} \frac{|x_0 + h| - |x_0|}{h} = -1.$$

So, in the point $x_0 = 0$, the left and right limits of the above difference fractions exist, but they are different, so that the function is not *differentiable*.

In the complex plane, the situation becomes more complicated. Let $D \subset \mathbb{C}$ be a domain, $f \colon D \to \mathbb{C}$ a function, and $z_0 \in D$. The function f is *differentiable* in point z_0, if there is a unique value for the limit:

$$\lim_{h \to 0} \frac{f(z_0 + h) - f(z_0)}{h} \overset{\text{def}}{=} f'(z_0).$$

The function f' is called the *derivative* of f. The quantity h is here a *complex number*: when computing the limit, we may approach the point x_0 from any direction, not just from the positive or negative direction of the real axis, but also from "above" or "below," and from all other directions.

A simple example of a non-differentiable function is the function $f(z) = |z|$ at the origin $z = 0$. To show this, it suffices to observe that the limits $h \uparrow 0$ and $h \downarrow 0$ along the real axis are different values. But additionally it may be observed ($h \in \mathbb{R}$) that

$$\lim_{h \downarrow 0} \frac{|z_0 + ih| - |z_0|}{ih} = -i \quad \text{and} \quad \lim_{h \uparrow 0} \frac{|z_0 + ih| - |z_0|}{ih} = i$$

at the origin $z_0 = 0$.

The function f is *analytic* in a domain D, if it is differentiable in every point $z \in D$. The function f is analytic in point z_0, if it is analytic in some neighbourhood of point z_0.

For analytic functions, the usual differentiation rules apply, the proofs of which run like in the real case:

$$(f + g)'(z) = f'(z) + g'(z) \qquad \text{(linearity)},$$
$$(fg)'(z) = f'(z)\,g(z) + f(z)\,g'(z) \qquad \text{(Leibniz product rule)},$$
$$\left(\frac{f}{g}\right)'(z) = \frac{f'(z)\,g(z) - f(z)\,g'(z)}{\big(g(z)\big)^2} \qquad \text{if } g(z) \neq 0,$$
$$(f \circ g)'(z) = f'\big(g(z)\big)\,g'(z) \qquad \text{(chain rule)}.$$

It holds especially, that the sum and product of analytic functions are analytic. Therefore, e.g., polynomials are analytic functions.

If $f(z)$ is an analytic function in a domain D, then also the derivatives of all orders exist in the domain D, and they are also analytic functions in the domain D.

The fraction of two polynomials $P(z)$ and $Q(z)$,

$$f(z) = \frac{P(z)}{Q(z)},$$

is called a *rational function*. A rational function is analytic except in those points where $Q(z) = 0$. Here we shall assume that the fraction has been simplified by eliminating the common factors of P and Q.

4.3   The Cauchy–Riemann conditions

Figure 4.2: Augustin-Louis Cauchy, painting by Jean Roller from around 1840.

Let us assume that the function $f\colon D \to \mathbb{C}$, $f(z) = u(x, y) + iv(x, y)$ is defined and continuous in the domain D. Write the analytic function using the real and imaginary parts

$$f(z) = u(x, y) + iv(x, y),$$

where $z = x + iy \in D$. From the differentiability in point z follow the *Cauchy[1]–Riemann conditions*:

$$\begin{aligned}
\frac{\partial}{\partial x} u(x, y) &= \frac{\partial}{\partial y} v(x, y), \\
\frac{\partial}{\partial y} u(x, y) &= -\frac{\partial}{\partial x} v(x, y),
\end{aligned} \tag{4.1}$$

i.e., if the function $f(z)$ exists in some neighbourhood of point $z = x + iy$ and is differentiable in point z, then the functions u and v have partial derivatives in the point z, and they satisfy the Cauchy–Riemann conditions, Equations 4.1. Proving these equations is not hard, and the proof is found in all textbooks of complex calculus.

[1] Augustin-Louis Cauchy (1789 – 1857), Figure 4.2, was a gifted French mathematician and pioneer of, among other things, complex analysis.

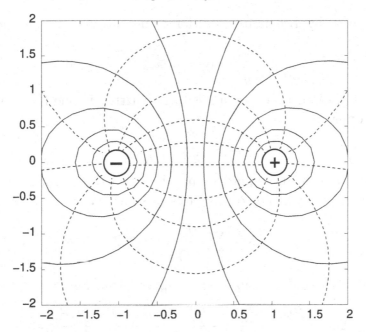

Figure 4.3: The equipotential curves (solid lines) and force lines (dashed lines) of a dipole field. The curves are everywhere mutually perpendicular. The field depicts the complex function $w = \mathrm{Ln}(z-1) - \mathrm{Ln}(z+1)$.

In particular, if $f(z)$ is analytic in the domain $D \subset \mathbb{C}$, then the partial derivatives exist and satisfy the Cauchy–Riemann conditions 4.1 for all $z \in D$. On the other hand, if $u(x,y)$ and $v(x,y)$ are functions of two real variables which have continuous partial derivatives with respect to the variables x, y satisfying the Cauchy–Riemann conditions 4.1 in the domain D, then the complex function $f(z) = u(x,y) + iv(x,y)$ is analytic in the domain D.

The Cauchy–Riemann conditions describe a property of analytic functions that can also be presented geometrically, and that is broadly useful in physics. The curves $u = $ constant and $v = $ constant are mutually *orthogonal*. As Figure 4.3 shows, one may interpret, e.g., v as the potential of some force field, in which case the curves $u = $ constant, which run orthogonally to the $v = $ constant curves, are running everywhere in the direction of the gradient of the potential, i.e., the force vector.

The situation depicted in the figure is two-dimensional; the sources in places -1 (negative) and $+1$ (positive) could be, e.g., electrostatically charged, infinitely long copper wires, extended in the direction perpendicular to the plane. Then, the drawn lines depict the electrostatic potential's equipotential curves and the dashed lines, the force lines of the electrostatic field.

Alternatively, one may think that the copper wires carry current — in opposite directions — and the drawn lines are now the force lines of the magnetic field generated.

From the Cauchy–Riemann conditions, one obtains still the *Laplace*[2] *equations*:

$$\frac{\partial^2 u}{\partial x^2} + \frac{\partial^2 u}{\partial y^2} = 0, \qquad \frac{\partial^2 v}{\partial x^2} + \frac{\partial^2 v}{\partial y^2} = 0. \tag{4.2}$$

The Laplace equation, Equation 4.2 is often written in the form $\Delta u = 0$ using the Laplace operator

$$\Delta = \frac{\partial^2}{\partial x^2} + \frac{\partial^2}{\partial y^2}. \tag{4.3}$$

The Laplace operator and equation may be defined also for a higher number of dimensions. If we call the co-ordinates of space $x_1, x_2, \ldots, x_n$:

$$\Delta \stackrel{\text{def}}{=} \sum_{j=1}^{n} \frac{\partial^2}{\partial x_j^2}, \quad n \geq 2. \tag{4.4}$$

For example, in three dimensions, the Laplace equation is

$$\Delta u = \frac{\partial^2 u}{\partial x^2} + \frac{\partial^2 u}{\partial y^2} + \frac{\partial^2 u}{\partial z^2} = \frac{\partial^2 u}{\partial x_1^2} + \frac{\partial^2 u}{\partial x_2^2} + \frac{\partial^2 u}{\partial x_3^2} = 0.$$

A real-valued function for which the Laplace equation holds is called a *harmonic* function. Such functions occur often in natural science and engineering: for example, the already mentioned electrostatic potential, but also the velocity vector of laminary (non-turbulent) air flow around an aircraft wing may be understood as the gradient of a potential function, see e.g., NASA, Joukowski Transformation. So, both the real and the imaginary parts of an analytic function are harmonic functions. Actually, this is the way to obtain all harmonic functions that exist in the complex plane. There is no corresponding result for higher dimensions.

4.4   Series expansions

Let $z_1, z_2, z_3, \ldots$ be a sequence of complex numbers. Use the notation

$$s_1 = z_1, \; s_2 = z_1 + z_2, \text{ and generally } s_n = z_1 + z_2 + \ldots + z_n.$$

Then the number s_n is called the nth partial sum of the *series*

$$\sum_{k=1}^{\infty} z_k = z_1 + z_2 + z_3 + \ldots$$

---

[2]Pierre-Simon Laplace (1749 – 1827) was an astronomer, mathematician, physicist and universal genius. He wrote, among other things, a textbook on celestial mechanics in five parts.

The series is called *convergent* if

$$\lim_{n \to \infty} s_n = s,$$

and then, the number $s \in \mathbb{C}$ is called the sum of the series. If the series does not converge, it is called *divergent*.

The complex series $\sum_{k=1}^{\infty} z_k$, where $z_k \in \mathbb{C}$, can be understood with the help of its real and imaginary parts. One obtains, with $z_k = x_k + iy_k$,

$$\sum_{k=1}^{\infty} z_k = \lim_{n \to \infty} \sum_{k=1}^{n} z_k = \sum_{k=1}^{\infty} x_k + i \sum_{k=1}^{\infty} y_k,$$

if both $\sum_{k=1}^{\infty} x_k$ and $\sum_{k=1}^{\infty} y_k$ converge. If the series $z_1 + z_2 + \ldots$ converges, then $\lim_{k \to \infty} z_k = 0$. Especially, if this is not the case, then the series diverges. Especially if the right-hand side series converges, then

$$\left| \sum_{n=1}^{\infty} z_n \right| \leq \sum_{n=1}^{\infty} |z_n|.$$

Example. As in the real case, we study the geometric series

$$\sum_{n=0}^{\infty} z^n = \frac{1}{1-z}, \text{ if } |z| < 1, \ z \in \mathbb{C}.$$

Based on the ratio test, this series converges on the open disc $|z| < 1$:

$$\sum_{n=1}^{\infty} z_n \text{ converges if } \lim_{n \to \infty} \left| \frac{z_{n+1}}{z_n} \right| < 1,$$

and diverges if

$$\lim_{n \to \infty} \left| \frac{z_{n+1}}{z_n} \right| > 1.$$

4.5 An analytic function as a power series

The most practically useful type of series is the *power series*. The power series is of the form

$$\sum_{n=0}^{\infty} a_n (z - z_0)^n, \tag{4.5}$$

where $z_0 \in \mathbb{C}$ is the *centre*, and the numbers $a_n \in \mathbb{C}$ are called the *coefficients* of the series.

The power series 4.5 converges either

○ in the whole complex plane (for all $z \in \mathbb{C}$),

○ only in point $z = z_0$, or

○ on some disc
$$D(z_0, R) = \{z : |z - z_0| < R\},$$

and possibly in some of the boundary points of the disc, but diverges outside the disc. The number R is called the *radius of convergence* of the series. In the first case we may think $R = \infty$, and in the second, that $R = 0$. The power series thus defines a function $f : D(z_0, R) \to \mathbb{C}$,

$$f(z) = \sum_{n=0}^{\infty} a_n (z - z_0)^n,$$

which is the *sum function* of the series. For example, the exponential function and the trigonometric functions may be expressed as power series which converge in the whole complex plane. The polynomial functions can be thought of as power series with a finite number of terms.

Can every analytic function be expressed as the sum function of a power series? There is no corresponding result in the real calculus. We may, e.g., look at the function $f : \mathbb{R} \to \mathbb{R}$,

$$f(x) = \begin{cases} e^{-1/x}, & x > 0, \\ 0, & x \leq 0. \end{cases}$$

Because $f^{(n)}(0) = 0$ for all n, the function's Taylor series is zero and does not describe the function at any $x > 0$. This can, however, not happen in the complex case. We obtain the following result:

Theorem 4.1 (Taylor expansion). *Let $f : D \to \mathbb{C}$ be an analytic function, and $z_0 \in D$. Then*

$$f(z) = \sum_{n=0}^{\infty} \frac{f^{(n)}(z_0)}{n!} (z - z_0)^n$$

for all $z \in D(z_0, R)$, in which R is the distance of point z_0 from the boundary ∂D of the domain D.

Remark. Here, R is the convergence radius of the series, if D is the maximal domain of definition of the function f.

Example. The series expansion

$$f(z) = \frac{1}{1 + z^2} = \sum_{n=0}^{\infty} (-1)^n z^{2n}$$

applies in the unit disc, when $|z| < 1$. The definition set of the function is $D = \mathbb{C} \setminus \{\pm i\}$ and the centre of expansion is $z_0 = 0$, so that $R = 1$. Also $R = |0 - (\pm i)|$, i.e., the distance of the origin from the boundary of D.

Remark. For real numbers, the function

$$f(x) = \frac{1}{1 + x^2}$$

is defined, but, also in the real-valued case, the series converges only, when $x \in (-1, 1)$.

4.6 Complex Taylor series

From the results obtained, it follows that from the real Taylor series, complex Taylor[3] series are obtained by substituting $x \mapsto z$, $z \in \mathbb{C}$.

Examples. The series

$$e^z = \sum_{n=0}^{\infty} \frac{1}{n!} z^n,$$

$$\sin z = \sum_{n=0}^{\infty} \frac{(-1)^n}{(2n+1)!} z^{2n+1},$$

$$\cos z = \sum_{n=0}^{\infty} \frac{(-1)^n}{(2n)!} z^{2n},$$

$$\mathrm{Ln}(1 + z) = \sum_{n=1}^{\infty} \frac{(-1)^{n+1}}{n} z^n, \quad |z| < 1.$$

Other examples of complex Taylor series are

$$(1 + z)^p = 1 + \sum_{n=1}^{\infty} \frac{p(p-1)(p-2)\cdots(p-n+1)}{n!} z^n, \quad |z| < 1,$$

and

$$f(z) = z e^{3z} = z \sum_{n=0}^{\infty} \frac{1}{n!} (3z)^n = \sum_{n=0}^{\infty} \frac{3^n}{n!} z^{n+1}, \quad z \in \mathbb{C}.$$

Remark. The function

$$f(z) = \frac{\sin z}{z}$$

is analytic where $z \neq 0$. As its Taylor series is obtained

$$\frac{\sin z}{z} = \sum_{n=0}^{\infty} \frac{(-1)^n}{(2n+1)!} z^{2n} = 1 - \frac{1}{3!} z^2 + \frac{1}{5!} z^4 - \cdots$$

when $z \neq 0$. We observe, however, that this series represents the analytic function throughout the complex plane, i.e., $R = \infty$. So, if we define $f(0) = 1$, we obtain an analytic function $f \colon \mathbb{C} \to \mathbb{C}$.

---

[3]Brook Taylor (1685–1731) was an English mathematician known mostly for the Taylor series expansion, Wikipedia, Taylor's theorem.

Exercises

Exercise 4 – 1:

Study the convergence of the following series:

a. $\displaystyle\sum_{n=1}^{\infty} \frac{1}{\sqrt{n}},$

b. $\displaystyle\sum_{n=1}^{\infty} \frac{i^n}{n}.$

Exercise 4 – 2:

If $f(z)$ and $\overline{f(z)}$ are both analytic functions in the domain D, show by applying the Cauchy–Riemann equations that $f(z)$ is constant in D.

Exercise 4 – 3:

Find such a constant a, that the function

$$f(x + iy) = x^2 - y^2 + x + i\,(axy + y)$$

is analytic.

Exercise 4 – 4:

Can a function be differentiable in some point z without being analytic in it? If yes, give an example.

Chapter 5

Conformal mappings

The mapping (function) $w = f(z)$ is said to be *conformal* if it preserves the angle between two intersecting, smooth paths (meaning it preserves both the magnitude and the orientation[1]). Here, the angle between two curves means the angle between their tangents at the point of intersection z_0.

Example. The mapping $z \mapsto z^2$ doubles the angles in point 0. So, it is not conformal in this point. Actually, it is conformal in all other points.

Theorem 5.1. *The analytic function $f \colon D \to f(D)$ is conformal in all those points $z \in D$ for which $f'(z) \neq 0$.*

Justification. Assume $z_0 \in D$ and $f'(z_0) \neq 0$. Study the path

$$C \colon z(t) = x(t) + iy(t)$$

in point $z_0 \in D$. If $z(t_0) = z_0$, then $z'(t_0)$ is the tangent of path C in point z_0. The image of path C is the composed mapping $w(t) = f(z(t))$. With the chain rule of differentiation we obtain $w'(t) = f'(z(t)) \, z'(t)$. Because the derivative of f in a chosen point is a complex number (as assumed, $f'(z_0) \neq 0$), we see that the mapping f changes in point z_0 the arguments of all paths by the same amount, i.e., preserves angles.

An important result concerning conformal mappings is the Riemann mapping theorem. This result says that there are many conformal mappings: in fact, every simply connected domain[2] D, which is not the whole complex plane, can be mapped conformally into an open unit disc $\mathbb{D} = \{z : |z| < 1\}$. If it is additionally assumed that $z_0 \in D$, $f(z_0) = 0$ and $|f'(z_0)| > 0$, then, based on Poincaré[3]'s result, the mapping is also unique.

---

[1]"Orientation" here means going either clockwise or counterclockwise when turning from the first path to the second.

[2]A domain D is called simply connected if, for all paths lying completely within the domain D, it holds that also the set contained within the path is fully inside the domain D. Intuitively this means, that the domain contains no "holes."

[3]Jules Henri Poincaré (1854−1912) was a French mathematician, physicist, engineer and philosopher of science. Poincaré has often been called the last human being who knew all of mathematics. Of his many achievements, the most famous is the general solution of the so-called "three-body problem," which had a significant effect on the birth of chaos theory.

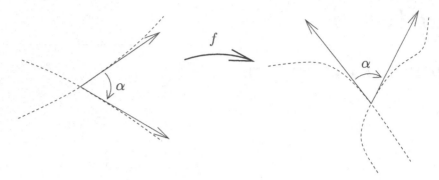

Figure 5.1: Conformal mapping.

The Riemann mapping theorem is, as a mathematical result, extremely powerful and even counterintuitive. Its proof can be found in more extensive textbooks of complex analysis such as Ahlfors (1966). Unfortunately, the result and its proof do not provide any obvious method or algorithm for finding the conformal mapping. Recently, however, numerical algorithms for finding conformal mappings have been the subject of active research, and for this purpose a number of software packages are nowadays available.

5.1 Helmert mappings or transformations

Helmert[4] mappings or transformations can be written in complex form thusly:

$$f(z) = \alpha z + \beta,$$

in which α and β are complex constants. The constant $\beta = a + ib$ is called the *translation vector*; the constant

$$\alpha = (1 + \mu)e^{i\theta}$$

is the complex scale. Here, μ is the *scale distortion* and θ is the *rotation angle*. In this so-called plane Helmert transformation, there are thus two complex-valued or four real-valued parameters. Mathematically, the Helmert transformation is the same as the complex affine transformation[5].

We may write the same equation also as a (real-valued) matric equation:

$$\begin{bmatrix} u(z) \\ v(z) \end{bmatrix} = (1 + \mu) \cdot \begin{bmatrix} \cos\theta & -\sin\theta \\ \sin\theta & \cos\theta \end{bmatrix} \cdot \begin{bmatrix} x \\ y \end{bmatrix} + \begin{bmatrix} a \\ b \end{bmatrix},$$

[4]Friedrich Robert Helmert (1843–1917) was a German geodesist, a gifted developer of theory and computational methods, and an author of textbooks.

[5]Note however that all affine transformations in $\mathbb{R}^2$ are not affine transformations in the sense of complex analysis.

where $f(z) = u(z) + iv(z)$ and $z = x + iy$.

Often encountered is the case where both the scale μ and the rotation angle θ, and possibly also the shifts a and b, are small. Then we may write:

$$\alpha = (1 + \mu)e^{i\theta} \approx 1 + \mu + i\theta$$

and $\sin\theta \approx \theta$ and $\cos\theta \approx 1$. The result is, considering that also $\mu\theta \approx 0$:

$$\left[\begin{array}{c} u(z) \\ v(z) \end{array} \right] = \left[\begin{array}{cc} 1 + \mu & -\theta \\ \theta & 1 + \mu \end{array} \right] \cdot \left[\begin{array}{c} x \\ y \end{array} \right] + \left[\begin{array}{c} a \\ b \end{array} \right]$$

or

$$\left[\begin{array}{c} u - x \\ v - y \end{array} \right] \approx \left[\begin{array}{cc} \mu & -\theta \\ \theta & \mu \end{array} \right] \cdot \left[\begin{array}{c} x_0 \\ y_0 \end{array} \right] + \left[\begin{array}{c} a \\ b \end{array} \right], \tag{5.1}$$

where the vector $\left[\begin{array}{cc} x_0 & y_0 \end{array} \right]^{\mathsf{T}}$ consists of *approximate values*: the 2×2 matrix in the formula contains only the *small* constants μ and θ. Equation 5.1 is useful because it describes the small change of co-ordinates in the case where the (u, v) co-ordinates and the (x, y) co-ordinates are close to each other. These cases are encountered often in geodesy, for example, co-ordinate frames that are in principle the same, are determined in the same way, but are nevertheless based on different sets of observations.

The two-dimensional Helmert transformation results always when *linearizing* an analytic function in the complex plane. This can be easily demonstrated using a Taylor expansion:

$$f(z) = f(z_0) + \left. \frac{df}{dz} \right|_{z=z_0} (z - z_0) + \ldots \quad \approx \alpha z + \beta,$$

if we define

$$\alpha = \left. \frac{df}{dz} \right|_{z=z_0}, \qquad \beta = f(z_0) - \alpha z_0.$$

This linearization is always valid in a sufficiently small neighbourhood of point z_0. How big or how small this neighbourhood will be depends on the behaviour of the function $f(z)$.

The two-dimensional Helmert transformation may be readily generalized to the three-dimensional transformation:

$$\left[\begin{array}{c} x' \\ y' \\ z' \end{array} \right] \approx \left[\begin{array}{ccc} 1 + \mu & \epsilon_z & -\epsilon_y \\ \epsilon_z & 1 + \mu & \epsilon_x \\ -\epsilon_y & \epsilon_x & 1 + \mu \end{array} \right] \cdot \left[\begin{array}{c} x \\ y \\ z \end{array} \right] + \left[\begin{array}{c} \Delta x \\ \Delta y \\ \Delta z \end{array} \right]. \tag{5.2}$$

Here are now *seven* parameters: three translation or shift components, $\Delta x, \Delta y$ and Δz, three rotation angles ϵ_x, ϵ_y and ϵ_z, as well as a scale distortion μ. Equation 5.2 is widely used for transforming co-ordinate material in satellite positioning, i.e., three-dimensional geocentric, datums into each other. Also this formula is valid when the transformation is "small": the co-ordinates are close to each other and both μ and the epsilons are small numbers.

Unfortunately, there is no direct equivalent to complex numbers for three-dimensional space. For this purpose, one has tried to use the *quaternions*. Quaternions are numbers $Q = a + ix + jy + kz$, with calculation rules:

$$ij = k, \qquad jk = i, \qquad ki = j,$$
$$ji = -k, \qquad kj = -i, \qquad ik = -j,$$
$$i^2 = j^2 = k^2 = -1.$$

Quaternions are like complex numbers, but not nearly as useful. Their inventor was Sir William Rowan Hamilton (1805–1865) of Dublin[6]. He was a famous mathematican and physicist who invented a new way to formalize Newton's classical mechanics, in which the path of an object is the same "shortest path" as in optics the path of a light ray. He was about a century ahead of his time, and not until the advent of quantum theory did it become clear, that a wave phenomenon guides the motion of also other particles than photons. Similarly also the general theory of relativity showed that the *world line* describing the motion of an object is a geodesic in space-time.

Also quaternions remained long a curiosity[7], until Paul Dirac[8] showed in 1928 that the matter wave of the electron is a *spinor field*. Spinors are closely related to quaternions. Paul Kustaanheimo[9] and Eduard L. Stiefel[10] again applied them in 1965 to celestial mechanics (Kustaanheimo and Stiefel, 1965).

---

[6]According to a story, Hamilton invented quaternions on Monday October 16, 1843, on a walk, and he carved the formulas into the stone of Broom (Brougham) Bridge close to Dublin. The carving has since vanished, but there is a memorial plaque on the location, Wikipedia, Broom Bridge, and it is the city's foremost place of pilgrimage for mathematicians right after the Guinness brewery (Wikipedia, William Sealy Gosset).

[7]However, Klein and Sommerfeld (1897, pages 55–68) mention them in their fundamental work on the motion of the gyroscope.

[8]Paul Adrien Maurice Dirac (1902–1984) was an English theoretical phyisicist. He developed a wave equation for the electron that is compatible with the special theory of relativity, and predicted, on the basis of this, the existence of the anti-electron or positron. He was considered a bit weird by colleagues and friends (Farmelo, 2011), and was, in light of today's understanding, probably afflicted with Asperger's syndrome (Silberman, 2015).

[9]Paul Edwin Kustaanheimo (1924–1997; –1935 Gustavsson) was a Finnish astronomer, who worked for a long time in Denmark.

[10]Eduard L. Stiefel (1909–1978) was a Swiss mathematician. He is also known as the developer of the conjugate-gradient method for efficiently solving large systems of linear equations.

And in very recent years, quaternions are being used also in three-dimensional simulation and modelling computations describing the motions of cameras — and, e.g., in video games.

5.2 Möbius transformations

Möbius[11] transformations (also called "fractional linear transformations") are transformations which, in the complex plane, may be defined by an equation of the form

$$f(z) = \frac{az + b}{cz + d}, \qquad (ad - bc \neq 0), \tag{5.3}$$

where a, b, c and d are complex (or real) constants. Because

$$f'(z) = \frac{a(cz + d) - c(az + b)}{(cz + d)^2} = \frac{ad - bc}{(cz + d)^2},$$

we see that $f'(z) \neq 0$ (and thus f is conformal) for all $z \in \mathbb{C}$. The inverse transformation of the Möbius transformation $w = f(z)$, i.e., the transformation $z = f^{-1}(w)$, is according to Equation 5.3 obtained from the formula

$$f^{-1}(w) = \frac{dw - b}{-cw + a},$$

from which we see that the inverse of a Möbius transformation is always a Möbius transformation.

Special Möbius transformations are:

Figure 5.2: August Ferdinand Möbius, as painted by Adolf Neumann.

o translations (shifts) $z \mapsto z + a$, $a \in \mathbb{C}$ constant,

o rotations $z \mapsto az$, $|a| = 1$,

o scalings $z \mapsto az$, $\operatorname{Arg} a = 0$,

o similarity transformations $z \mapsto az + b$, $a \neq 0$,

o mirroring through the unit circle (inversion) $z \mapsto \dfrac{1}{z}$.

---

[11]August Ferdinand Möbius (1790 – 1868), see Figure 5.2, was a German mathematician and astronomer, known as a pioneer of topology.

Actually all Möbius transformations may be described as combinations of finitely many such mappings. This can be easily proven: start from the inversion (mirroring):

$$f(z) = w = \frac{1}{z},$$

and add *scaling* and *rotation r*:

$$w = r \cdot \frac{1}{z} = \frac{r}{z}.$$

Add translation before and after:

$$w + p = \frac{r}{z + q},$$

i.e.,

$$f(z) = w = \frac{r}{z + q} - p = \frac{r}{z + q} - p \cdot \frac{z + q}{z + q} = \frac{r + p(z + q)}{z + q} = \frac{pz + (r + pq)}{z + q},$$

which is of the form 5.3, if we consider that the formula may be normalized in such a way that the constant $c = 1$ vanishes. From this we see that the Möbius transformation has three complex parameters, e.g., p, q, r or a, b, d when $c = 1$. Conversely, we may define that Möbius transformations are all transformations of the above form.

Remark. According to the so-called generalized Liouville[12] theorem, each conformal mapping in space $f \colon D \to \mathbb{R}^3$ (in which $D \subset \mathbb{R}^3$ is a domain), that is three times continuously differentiable (i.e., a C^3 mapping), is the restriction of a Möbius transformation (Liouville, 1850). Generally speaking, the situation in the case of $\mathbb{R}^3$ is, very differently from the case of two dimensions, that, in addition to every Möbius transformation being conformal, *also all conformal mappings* (satisfying certain requirements) *are Möbius transformations*! So, in three dimensions, conformality is a much more constraining requirement than in the plane. This result can be directly generalized to all Euclidean spaces $\mathbb{R}^n$, where $n \geq 3$.

The assumption of three times differentiability is not essential. For C^1 mappings, the result in question was demonstrated by Hartman (1958), and for so-called weakly differential mappings, the result was demonstrated by

---

[12]Joseph Liouville (1809–1882) was a French mathematician who brought many fields within mathematics forward.

Gehring[13] (1962) and Reshetnyak[14] (1967). A simple proof for C^4 mappings was presented by Rolf Nevanlinna[15] (1960).

An example of a Möbius transformation in three-dimensional space is the stereographic projection described elsewhere. On two-dimensional surfaces in three-dimensional space, it is however possible to define also other conformal mappings.

The image of a straight line or circle under a Möbius transformation is always a straight line or a circle. The hardest part of the proof of this is showing that the *inversion* $f(z) = \frac{1}{z}$ maps circles to circles. Because every Möbius transformation can always be formed by combining a finite number of inversions and linear transformations $f(z) = az + b$ (which trivially map circles to circles), the result follows.

Let us prove that the inversion maps circles to circles. Let us give the proof for the mapping

$$w = f(z) = \frac{1}{z}.$$

Note that the complex conjugate trivially maps circles to circles (even if they are of opposite directionality). We see immediately that

$$|w| = \frac{1}{|z|} \quad \text{and} \quad \text{Arg}\, w = \text{Arg}\, z.$$

See Figure 5.3. First we prove that the following triangles are similar: $\triangle OPB \sim \triangle OB'P'$ and $\triangle OPA \sim OA'P'$ — when the triangles have a common angle, e.g., θ_1, and the ratio of two sides is the same, e.g., $OB : OP = OP' : OB'$. From this follows, that β_1 and β_2 have both been drawn correctly in the two places. It also follows that

$$\alpha_1 = \beta_1 - \theta_1,$$
$$\alpha_2 = \beta_2 - \theta_2.$$

Let it also be given that the small circle closer to the origin is indeed a circle. If the location of points A and B is constant, then the angle $\beta = \beta_1 + \beta_2$ is a *constant*, independently from where point P lies on the circle.

But then, $\alpha = \alpha_1 + \alpha_2 = \beta_1 + \beta_2 - \theta_1 - \theta_2$ is constant, and also the set of possible locations of point P' *must be a circle*. In this proof it is assumed, that

---

[13] Frederick William Gehring (1925–2012) was an American mathematician and researcher of geometric function theory. Gehring is known especially for his results on the theory of quasi-conformal mappings.

[14] Yurii Grigorievich Reshetnyak (1929–) is a Russian mathematician, academician and researcher of geometric function theory.

[15] Rolf Herman Nevanlinna (1895–1980) was a Finnish mathematician known especially for his studies related to complex analysis.

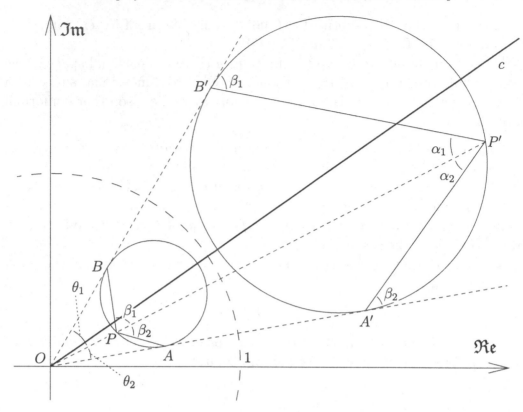

Figure 5.3: Inversion maps circles into circles.

the origin O is not located on the circumference of the circle to be mapped. It this were the case, the result would be a straight line (which can then be understood as a circle passing through the infinity point).

By using a Möbius transformation, one may always change any three points in the plane into other points. In this way, one may construct conformal mappings that, e.g., change a given disc into a semi-plane, or the reverse.

Theorem 5.2. *If given are three separate points z_1, z_2, z_3 in the complex plane, one may always map them (preserving their order) into the triple w_1, w_2, w_3 using a unique Möbius map, which is found by solving $w(z)$ from the equation*

$$\frac{(w - w_1)(w_2 - w_3)}{(w - w_3)(w_2 - w_1)} = \frac{(z - z_1)(z_2 - z_3)}{(z - z_3)(z_2 - z_1)}.$$

If any of the points is an infinity point, the formula may be interpreted as a limit.

Remark. If z_1, z_2, z_3 and z_4 are complex numbers, the expression

$$(z_1, z_2, z_3, z_4) = \frac{(z_1 - z_2)(z_3 - z_4)}{(z_1 - z_4)(z_3 - z_2)}.$$

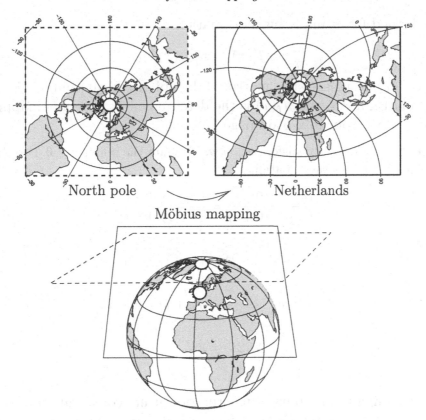

North pole

Netherlands

Möbius mapping

Figure 5.4: Möbius transformation as a combination of two stereographic map projections. This is exact only on a spherical Earth! Here we have used as an example the projection in its base orientation, in which the North Pole is the projection centre, as well as the stereographic projection used for topographic maps in the Netherlands, where the spire of the old Church of Our Lady in Amersfoort acts as the projection centre.

is called a *cross-ratio*. Theorem 5.2 may be understood to say that a mapping is a Möbius transformation if and only if it preserves the cross-ratio.

A corresponding result applies also in three- and higher-dimensional spaces. In that case, however, one should first define a generalization of the cross-ratio, the *absolute ratio*, by the equation

$$\|x_1, x_2, x_3, x_4\| = \frac{\|x_1 - x_2\| \|x_3 - x_4\|}{\|x_1 - x_4\| \|x_3 - x_2\|},$$

in which $x_1, x_2, x_3, x_4 \in \mathbb{R}^n$, $n \geq 2$. Because the conservation of the absolute ratio does not guarantee that the mapping preserves orientation (in other words, is not a mirroring) this must be separately assumed. Also in higher-dimensional spaces, Möbius transformations are conformal. The conformality of mapping f

may be defined for example by the equation

$$\|f'(x)\|^n = J_f(x), \tag{5.4}$$

in which $f'(x)$ designates the matrix of Jacobi of the mapping f, $\|\cdot\|$ is the ordinary matrix norm, and $J_f(x)$ is the determinant of Jacobi of mapping f in the point x. See Appendix F on page 255.

Example. Seek a Möbius transformation which maps the points $-1, 0, 1$ to the points $-i, 0, i$. Substitute into the equation

$$\frac{(w+i)(0-i)}{(w-i)(0+i)} = \frac{(z+1)(0-1)}{(z-1)(0+1)}.$$

We obtain

$$(w+i)(z-1) = (w-i)(z+1)$$

and thus

$$wz - w + iz - i = wz + w - iz - i.$$

So

$$2iz = 2w,$$

i.e., $f(z) = iz$.

Remark. Defining the Möbius transformation using conservation of the cross- or absolute ratio is useful also because with it, it is easy to see, that Möbius transformations form a *group*: the combination of two successive Möbius transformations is always a Möbius transformation, the identity transformation (which maps every point onto itself) is a Möbius transformation, and the inverse of a Möbius transformation is also a Möbius transformation. This again may be used, e.g., in proving that the combination of two stereographic projections is a Möbius transformation in the plane. All mappings needed are restrictions of Möbius transformations in $\mathbb{R}^3$.

5.3   Möbius transformations and the stereographic projection

As already noted, in a Möbius transformation, a circle will always map to a circle or a straight line, and a straight line always to a circle or straight line. If we call circles and straight lines together by the name "general circle," we may say that a Möbius transformation maps general circles to general circles. A straight line may be considered a circle of infinite radius. In that case, any set of three points defines a general circle.

Also the *stereographic projection* from the sphere to the plane has the property, that a circle on the sphere — either a great circle or an ordinary

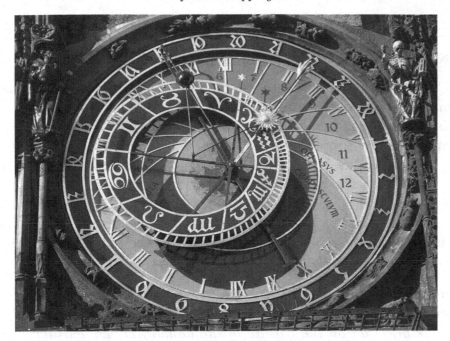

Figure 5.5: The clock face of the famous bell tower in Prague. The construction is based on the stereographic projection (image Wikipedia, Prague bell tower). By the same Old Town Square is also found the Týn church, the place of burial of Tycho Brahe.

circle — maps to a general circle in the projection plane. The proof is given in Appendix A.

In fact, the stereographic projection is a restriction of the three-dimensional inversion: the inversion centre is the projection centre and the radius of the inversion sphere is $2R$, twice the radius of the sphere being projected itself.

A Möbius transformation in the complex plane can always be written as a combination of two stereographic projections: the forward projection using one projection centre, and the backward projection using another one. In Figure 5.4 we try to show this graphically: the chosen projection centres are the North Pole and the Netherlands.

The stereographic projection is handy precisely because it maps circles to (general) circles. Projection figures may be drawn using the tools of classical geometry, ruler and compasses. Therefore, it has been used through the ages, e.g., in designing the *astrolabe*[16].

A beautiful astrolabe is shown in Figure 5.5.

---

[16]"Astrolabe" (Lat. *astrolabius*) is an astronomical observation instrument, originally developed by the Arabs in the middle ages, consisting of generally four parts turning about a common axis. A similar device, intended only to visualize the motions of the heavens, is also called "astrolabe."

5.4   Numerical conformal mappings

Let $D \subsetneq \mathbb{C}$ be a simply connected domain. Based on the Riemann mapping theorem, the domain D may be mapped conformally onto the unit disc $\mathbb{D}$, but this theorem does not help find any formula or other representation for the mapping in question. If the domain D is, however, somewhat simple in shape — like a polygon — then the conformal mapping may be approximated numerically. Often it is useful to assume, that the domain D to be mapped is a Jordan domain. This means that the boundary of domain D is a Jordan curve, i.e., a closed curve which does not intersect itself except of course in the end points. In that case, one may study values for the conformal mapping also on the boundary of the domain. One obtains the following result:

Theorem 5.3 (Carathéodory[17]– Osgood[18]–Taylor[19]). *Let D be a Jordan domain and $f \colon D \to \mathbb{D}$ a conformal mapping. Then f may be extended to a continuous bijection on the closure of the domain D, $\overline{D}$. Furthermore, any three points on the boundary of the domain D may be mapped to any three points placed in the same order on the boundary of the unit disc $\mathbb{D}$.*

Using this result, we may define the problem of finding the conformal mapping such that if given is a Jordan area D and three points z_1, z_2, z_3 on the boundary of the domain (moving in counter-clockwise direction), then we search for the mapping that takes the domain D to the unit disc and the given points, e.g., to points $-1, -i, 1$. The mapping thus obtained is unique.

Often it makes sense to take as the domain of definition the rectangle $R_h = [0, 1] \times [0, h]$, where the number $h > 0$ is called also the *conformal module* of the *generalized quadrangle* defined by the domain D and the four boundary points z_1, z_2, z_3, z_4. Then, the domain D is mapped onto the rectangle R_h and the points to be mapped go to its corner points. The quadrangle's conformal module and the conformal mapping thus obtained are unique.

---

[17]Constantin Carathéodory (1873–1950) was a Greek mathematician known especially for his results in real calculus, variational calculus and measure theory. He published a proof of this result in 1913 independently from the work of Osgood and Taylor.

[18]William Fogg Osgood (1864–1943) was an American mathematician known for his contributions to complex analysis and mathematical physics. He conjectured this result in 1903, but was only able to fully verify it in 1913 with his former student Taylor, independently from the result of Carathéodory published in the same year.

[19]Edson Homer Taylor (1874–1958) was an American mathematician who obtained his doctoral degree from Harvard University in 1909 under supervision of W. F. Osgood. Besides his role in the proof of this result, he is known for his contributions to teacher training and as an author of several popular mathematics textbooks.

An extremely useful MATLAB package for finding numerical conformal mappings is the Schwarz–Christoffel Toolbox[20] developed by Tobin Driscoll[21]. The logic of the package is based on numerically solving the Schwarz–Christoffel equation, the theoretical foundations of which are presented in the book by Driscoll and Trefethen (2002). It works in situations where the domain D to be mapped is a polygon. Other possible algorithms are, e.g., the Zipper algorithm, see Marshall (maintained), developed by Donald Marshall[22], as well as the conjugate function method developed by this book's author (Hakula et al., 2013). Numerical conformal mappings may also be studied for the case of Riemann surfaces, to be presented later in this book, for which the currently most popular method is based on so-called circle packing (Stephenson, 2005).

Exercises

Exercise 5 – 1:

Prove that the general similarity transformation

$$w = \alpha z + \beta.$$

maps circles to circles,

   a. algebraically, and

   b. geometrically.

Exercise 5 – 2:

Let f be the function $f(z) = e^z$, $z \in \mathbb{C}$.

   a. Express f in the form $f = u + iv$.

   b. Show that f is analytic in the whole complex plane.

   c. Give an example of such points z_1 and z_2, $z_1 \neq z_2$, that $f(z_1) = f(z_2)$.

---

[20]The package with instructions may be downloaded for free from Driscoll (maintained)'s web site.

[21]Tobin A. Driscoll is an American mathematician working as a professor at the University of Delaware. He is known for his contributions in numerical analysis.

[22]Donald E. Marshall is an American mathematician working as a professor at the University of Washington. He is known as an expert in complex analysis and algorithms related to it.

Exercise 5 – 3:

Let

$$f(z) = \frac{1+z}{1-z}.$$

    a. Show that f is conformal everywhere $z \neq 1$.

    b. Determine the image set of the unit circle $\{z : |z| = 1\}$ (i.e., the set $\{f(z) : |z| = 1, z \neq 1\}$).

Exercise 5 – 4:

Find a Möbius mapping that carries the points $-1, 0$ and 1 to the points $-i, 1$ and i.

Exercise 5 – 5:

Find a Möbius mapping that carries the points $0, -i$ and i to the points $-1, 0$ and ∞.

Exercise 5 – 6:

Let $T_a : \mathbb{R}^3 \to \mathbb{R}^3$ be a Möbius mapping that preserves the surface of the unit sphere $S^2 = \{x \in \mathbb{R}^3 : \|x\| = 1\}$ and maps the point a, $\|a\| < 1$ to the origin. What does the mapping T_a do to the antipodal point of a, the point $a^* = \dfrac{a}{\|a\|^2}$?

Hint: use the absolute ratio.

Exercise 5 – 7:

Show that a Möbius mapping in three-dimensional space carries a spherical surface $S^2(y, r) = \{x \in \mathbb{R}^3 : \|y - x\| = r\}$ to a spherical surface or a plane. Hint: also here one may use the absolute ratio. It is worthwhile to first normalize the situation with a similarity mapping.

Chapter 6

Transversal Mercator projections

Globally, the commonly used map projections for general or topographic maps have been *conformal* projections. In Finland, only transversal Mercator projections have been used. Historically, the Gauss–Krüger projection has been used, and as a novelty, also the Universal Transverse Mercator (UTM) projection is now in use in Finland. The UTM projection is a variant of Gauss–Krüger which has been globally already in widespread use. It is the standard map projection for the United States, suitable for global use, and also used by NATO.

6.1 Map projections used in Finland

Finland's history of map projection choices is rather typical, and about as messy as in many other countries. It is here presented as an illustrative case.

6.1.1 The historical Gauss–Krüger and KKJ

In Finland, the traditionally used map projection has been Gauss–Krüger with a zone width of 3°. The name of the full system was KKJ[1] (Finnish National Map Grid Co-ordinate System), and it was created in 1970 (Parm, 1988). Gauss–Krüger is a *transversal* cylinder projection, meaning that the "equator" of the projection is a chosen meridian of the system, the so-called central meridian. In Finland these meridians have been chosen at three-degree intervals, in order to be able to map the Finnish area in six different *projection zones*. See Table 6.1.

The idea is, that within every zone the scale distortion of the projection would remain as small as possible. For example, in zone 3, the longitude of the central meridian is 27° East of Greenwich, the area of Finland between longitudes 25.5° and 28.5° is mapped.

The scale distortion μ grows quadratically with distance from the central meridian according to the following equation:

$$\mu = 12.29 \cdot 10^{-15}\text{m}^{-2} \cdot (E - E_0)^2, \tag{6.1}$$

---

[1]Acronym written in all caps according to National Land Survey of Finland standardization recommendations, JUHTA (2016a).

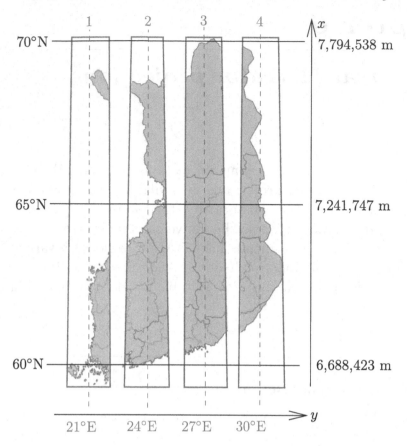

Figure 6.1: The four projection zones used in the Finnish KKJ system's Gauss–Krüger projection. In this figure, the zones 0 and 5 have been left out.

In Equation 6.1, E is "Easting," the metric co-ordinate in the East direction — E_0 again is the so-called "false Easting," in the case of Finland $500\,000\,\mathrm{m}$, the Easting of the central meridian. In this way, the Easting is always positive, which especially before the computer age was important in order to avoid

Table 6.1: Zones of the Finnish KKJ's Gauss–Krüger projection.

Zone	Central Meridian
0	18°
1	21°
2	24°
3	27°
4	30°
5	33°

confusion and clerical errors, as humans are better at visualizing numbers without algebraic signs.

We may calculate how large the scale distortion may become in Finland; as the semi-width of a projection zone is 1.5°, which at the latitude of Helsinki, 60°, corresponds to 83 kilometres, the distortion according to the above Equation 6.1 amounts to 0.000 085, or 85 ppm.

The above division in projection zones was in use for large-scale maps, i.e., maps on which a small area was depicted in great detail. Note that if the scale is $1 : m$, where m is a large number, a large scale means that m is small, for example, the scale $1 : 25\,000$ of a topographic map is a large scale. The scale 1:5000 of a zoning map is larger still. The example already shows what this is about: if a map is used for building and land-use planning, it is important that the scale is correct.

The *map-sheet division* of topographic maps is based, in the KKJ, also on this division in zones. Maps are drawn according to the plane co-ordinates x, y of the projection used in the zone.

For *small-scale* maps again the Gauss–Krüger projection was used for a central meridian of 27° for mapping all of Finland. For example, the scale of the GT road maps is $1 : 200\,000$. For these maps, a precisely correct scale factor is not critical, as already the difference between summer and winter tyres can amount to more than the effect of the map projection!

We can still make the following technical remarks on the projection of Finnish grid co-ordinate system:

o The reference ellipsoid used was the International (Hayford) ellipsoid of 1924, the starting point or *datum* was the triangulation point Simpsiö (number 90), the ED50 (European Datum 1950) co-ordinates and Bomford[2] geoid height (Bomford, 1963) were kept fixed.

o The co-ordinates obtained from ED50, (x, y), were *transformed* still inside the map plane using a four-parameter similarity transformation. The intent of this transformation was to produce KKJ co-ordinates which were close to the co-ordinates of the Helsinki System VVJ ("Old State System") already in widespread use. See Ollikainen (1993). The transformation formula used was:

$$\begin{bmatrix} x \\ y \end{bmatrix}_{KKJ} = \begin{bmatrix} 1.000\,000\,75 & -0.000\,004\,39 \\ 0.000\,004\,39 & 1.000\,000\,75 \end{bmatrix} \begin{bmatrix} x \\ y \end{bmatrix}_{ED50} + \begin{bmatrix} -61.571\,\mathrm{m} \\ 95.693\,\mathrm{m} \end{bmatrix}.$$

These details already fall under the subject of *realization* of a co-ordinate reference system, see Chapter 10 on page 127. Here they are mentioned as

---

[2]Guy Bomford (1899 – 1996) was a British officer and geodesist.

examples of the circumstance that there are many other decisions to be taken in connection to the precise mapping of the Earth's surface than just the choice of map projection.

6.1.2   Modern map projections Gauss–Krüger and UTM

Finland's modern map projection solution differs considerably from the old one. The whole solution is described in the Recommendation for Public Administration (JHS) number 197 (JUHTA, 2016b). The solution is based on the GRS80 reference ellipsoid, and the new *datum*, called EUREF-FIN, has been realized by GPS measurements. It is based on four datum points: the continuously operating GPS stations Metsähovi, Vaasa, Joensuu and Sodankylä (Ollikainen et al., 2000).

A more detailed discussion of this issue of realizing a modern geodetic datum and computing precise co-ordinates is found in the chapter on co-ordinate reference systems, Chapter 10 on page 127.

Also in the new solution, different choices have been made for large-scale and small-scale maps. For small-scale maps, the UTM projection is chosen, which is also globally extensively used — e.g., in the United States. It differs from Gauss–Krüger firstly in the scale on the central meridian not being 1.0000 but 0.9996. Compared to Equation 6.1 on page 67 we thus have

$$\mu = -0.0004 + 12.29 \cdot 10^{-15} \mathrm{m}^{-2} \cdot (E - E_0)^2 .$$

The other relevant difference is that the UTM system is global and the whole surface of the Earth has been divided into 60 projection zones, each 6° wide. The numbering of the zones starts at the date line and runs Eastward: zone 1 is 180°W − 172°W, zone 31 is thus 0° − 6°E, and zone 35 is 24°E − 30°E, with a central meridian of 27°.

For a visualization of both projection types, see Figure 6.2.

With a zone width of six degrees, the maximum distance from the central meridian is already on the equator over 300 km, and in Finland, the maximum distance from the 27° central meridian is in any case in excess of 400 km, and we already obtain for Gauss–Krüger scale distortions of order 1500 ppm! For construction work, zoning, and the cadastre, such distortions are simply not acceptable. But on small-scale maps, the scale pre-distortion of −400 ppm of the UTM projection will limit the maximum distortion for broad zones to something a bit more reasonable.

Therefore, the following choices have been made:

○ For small-scale maps, as well as for topographical maps, the UTM projection for central meridian 27° (according to UTM terminology, zone 35) is used *for all of Finland*. The name of this map co-ordinate frame is ETRS-TM35FIN. Also the map-sheet division is based on this.

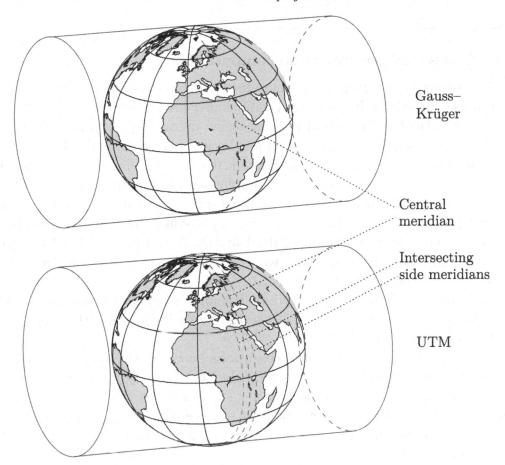

Gauss–
Krüger

Central
meridian

Intersecting
side meridians

UTM

Figure 6.2: The Gauss–Krüger and UTM projections. The UTM projection may be seen as an intersecting cylinder projection, which has two circles that are true of scale, on both sides of the central meridian.

However, in those parts of Finland in which another central meridian would have been more appropriate (like central meridian 21°, UTM zone 34), the corresponding co-ordinate grid is printed on the map in red (JUHTA, 2016b, JHS197).

o For large-scale maps, which are used in zoning and for the cadastre, the Gauss–Krüger projection continues to be used. However, the zone width here is just one degree: the naming is ETRS-GKn, in which n is the longitude of the central meridian in degrees. In this way, the scale distortions are even smaller still than in the old KKJ.

6.2 Gauss–Krüger projection computations

A good practical example of a map projection is the Gauss–Krüger projection in use in Finland. It is a transversal cylinder projection which is *conformal.*

The practical realization of the projection on an ellipsoid of revolution is not quite as simple as it would be on a spherical planet. Already the ordinary Mercator projection is clearly more complicated when realized on the ellipsoid. The transversality of the projection complicates matters further. There is no closed solution. Traditionally, the circumstance has been exploited that the flattening f is a small number, and a series expansion in this flattening has been sought which converges rapidly. Gifted mathematical geodesists of the 19th and 20th centuries (like Gauss and Krüger[3], as well as the Finnish Hirvonen[4]) worked on this problem (Hirvonen, 1972).

Also here, we shall present projection formulas as series expansions. But, unlike what has been traditionally the practice, we shall determine the coefficients of the expansion *numerically.*

We proceed as follows: exploit conditions based on the projection having to be conformal, and that it must map the central meridian correctly. First we choose a suitable *base projection*, e.g., the ordinary Mercator, for which the equations are simple. So, we map the surface of the Earth ellipsoid to the ordinary Mercator map plane, using Equations 2.7 on page 23, but applied to the zero meridian λ_0:

$$v = \lambda - \lambda_0, \qquad u = \int_0^\varphi \frac{M(\varphi')}{p(\varphi')} d\varphi'.$$

Next, we construct, in the Mercator plane, an *analytic mapping*

$$u + iv \;\mapsto\; x + iy,$$

of which one property is that x is of the correct length along the central meridian $\lambda - \lambda_0 = y = v = 0$:

$$dx = M(\varphi)\, d\varphi,$$

i.e.,

$$x = \int_0^\varphi M(\varphi')\, d\varphi'. \tag{6.2}$$

Furthermore, we have on the central meridian

$$y = 0.$$

---

[3]Johann Heinrich Louis Krüger (1857–1923) was a German mathematician and geodesist.

[4]Reino Antero Hirvonen (1908–1989) was a Finnish geodesist and student of the figure and gravity field of the Earth.

Table 6.2: Results of Gauss–Krüger test calculations. The number of terms means the number of coefficients needed in the polynomial expansion, to obtain a truncation error of under ±1 mm in the Gauss–Krüger co-ordinates computed.

Reference latitude (°)	Latitude interval (°)	Support pt. number	Computation pt. φ (°)	$\Delta\lambda$ (°)	No. of terms
1	1 – 20	20	19.333	10.0	12
61	61 – 80	20	79.333	20.0	16
70	61 – 80	20	79.333	20.0	15
70	61 – 80	19	79.333	20.0	15
65	61 – 70	19	60.333	10.0	14

Now we have derived a *boundary-value problem*, in which the sought-after complex map co-ordinate $z \stackrel{\text{def}}{=} x + iy$ is given as a function of the map co-ordinate of the base projection, $w \stackrel{\text{def}}{=} u + iv$, on the boundary $y = v = 0$, i.e., the real axis. The problem is determining the function *on the whole complex plane*. See Figure 6.3.

6.2.1 Analyticity and the Cauchy–Riemann conditions

In complex analysis one speaks of *analytic functions*. An analytic function is a mapping

$$z = f(w)$$

which is differentiable in some domain in the complex plane. It turns out that an analytic function is differentiable not just once, but infinitely many times. Then, the Cauchy–Riemann conditions 4.1 on page 46 apply:

$$\frac{\partial x}{\partial u} = \frac{\partial y}{\partial v}, \qquad \frac{\partial x}{\partial v} = -\frac{\partial y}{\partial u},$$

in which $z = x + iy$ and $w = u + iv$.

This means that a small vector $\begin{bmatrix} du & dv \end{bmatrix}^{\mathsf{T}}$ maps to a vector $\begin{bmatrix} dx & dy \end{bmatrix}^{\mathsf{T}}$ according to the following formula:

$$\begin{bmatrix} dx \\ dy \end{bmatrix} = \begin{bmatrix} a & -b \\ b & a \end{bmatrix} \begin{bmatrix} du \\ dv \end{bmatrix} = K \begin{bmatrix} \cos\theta & -\sin\theta \\ \sin\theta & \cos\theta \end{bmatrix} \begin{bmatrix} du \\ dv \end{bmatrix},$$

precisely as the Cauchy–Riemann conditions require.

From this we see that the mapping of local vectors $(du, dv) \mapsto (dx, dy)$ is a *scaling and rotation*. In this way we obtain a *conformal mapping*.

Almost all functions familiar from school math are analytic in some sub-domain of the complex plane, and many arithmetic operations preserve the analyticity of a function. See Section 4.2 on page 45.

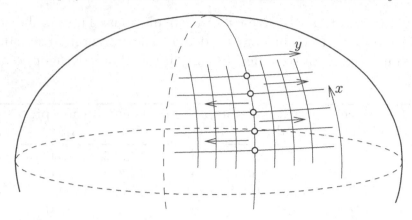

Figure 6.3: The Gauss–Krüger-projection as a boundary-value problem. The boundary values on the central meridian are marked with balls and the direction of integration with arrows.

6.2.2   Progress of the computation

Let us try as the general solution a *series expansion*:

$$z = a_0 + a_1 w + a_2 w^2 + a_3 w^3 + \ldots = \sum_{k=0}^{\infty} a_k w^k. \qquad (6.3)$$

Here we define for generality

$$z \overset{\text{def}}{=} (x - x_0) + iy, \qquad w \overset{\text{def}}{=} (u - u_0) + iv.$$

The values

$$x_0 \overset{\text{def}}{=} \int_0^{\varphi_0} M(\varphi)d\varphi, \qquad u_0 \overset{\text{def}}{=} \int_0^{\varphi_0} \frac{M(\varphi)}{p(\varphi)} d\varphi$$

have been chosen to be compatible with a suitable *reference latitude* φ_0. These series expansions thus are meant to be used only in a *relatively small domain*, not over the whole projection zone. In this way, also the number of terms needed remains smaller.

The meridian conditions 6.2 now form a set of *observation equations*:

$$x_i = a_0 + a_1 u_i + a_2 u_i^2 + a_3 u_i^3 + \ldots,$$

from which the coefficients a_k can be solved, when a sufficient number of support points (u_i, x_i) has been given (the balls in Figure 6.3). In Section 9.5 on page 118 we will explain how from this system of observation equations we may compute a least-squares solution: the unknowns (in Equation 9.7 on page 120, the elements of the vector $\widehat{x}$) in this problem are the coefficients a_k, the observables (elements of vector ℓ) the x co-ordinates of the meridian points, x_i (as the y co-ordinates all vanish). The design matrix $\mathbf{A}$ is made up of powers of the co-ordinates u_i.

Applying the coefficients a_k found to Equation 6.3 yields the solution in the whole complex plane[5]. The solution may be compacted into the following form that more suitable for computation, so-called Clenshaw[6] summation, see Tscherning and Poder (1982):

$$z = a_0 + w\left(a_1 + w\left(a_2 + w\left(a_3 + w\left(\cdots + wa_n\right)\right)\right)\right).$$

The sequence of computation is multiplication–summation–multiplication–summation... the powers of w do not need to be computed separately. Remember that also the intermediate results are complex numbers!

Fortunately, complex numbers are included in all popular programming languages, either as an integral part (MATLAB, Fortran) or as a standard library (C++).

In Table 6.2 some results obtained by the approach described here are given.

From the same equation both the scale and the meridian convergence are also easily obtained. As already proven, *scaling and rotation* are conformal mappings in the complex plane. When the equation of the mapping is 6.3, then its version expressed in differential notation is

$$\Delta z = \frac{dz}{dw}\Delta w, \tag{6.4}$$

in which

$$\frac{dz}{dw} = a_1 + 2a_2 w + 3a_3 w^2 + 4a_4 w^4 + \cdots = \sum_{k=1}^{\infty} k a_k w^{k-1}.$$

Also this may easily be written into the Clenshaw form.

Now, the complex Equation 6.4 can be written in real-valued matric form

$$\begin{bmatrix} dx \\ dy \end{bmatrix} = \begin{bmatrix} \mathfrak{Re}\left\{\frac{dz}{dw}\right\} & -\mathfrak{Im}\left\{\frac{dz}{dw}\right\} \\ \mathfrak{Im}\left\{\frac{dz}{dw}\right\} & \mathfrak{Re}\left\{\frac{dz}{dw}\right\} \end{bmatrix} \begin{bmatrix} du \\ dv \end{bmatrix},$$

in which $dz = \begin{bmatrix} dx & dy \end{bmatrix}^{\mathsf{T}}$, $dw = \begin{bmatrix} du & dv \end{bmatrix}^{\mathsf{T}}$,

$$\mathfrak{Re}\left\{\frac{dz}{dw}\right\} = \frac{\partial x}{\partial u} = \frac{\partial y}{\partial w} = K\cos\theta, \quad \mathfrak{Im}\left\{\frac{dz}{dw}\right\} = \frac{\partial y}{\partial u} = -\frac{\partial x}{\partial v} = K\sin\theta,$$

from which K, the *scale*, and θ, the *meridian convergence*, may be solved.

---

[5]The solution is unique. Conformity and the meridian condition together completely determine the solution.

[6]Charles William Clenshaw (1926–2004) was an English mathematician and pioneer of numerical computation. His research group developed those efficient algorithms for computing standard functions which are used, e.g., in function calculators, and which continue to be used in the numerical libraries of today's computers.

Remark. The choice of the classical Mercator projection as the starting projection is not the only alternative. One could probably reduce the number of terms in the series expansion 6.3 by using the Lambert conformal conical projection for reference latitude φ_0. One may, however, reflect on whether the achieved saving in computing work is worth the effort. A greater benefit is obtained by the system of equations formed for solving the problem having a better *condition number*, i.e., a lower *sensitivity to perturbations*, and with that, the numerical precision of the solution would improve.

Chapter 7

Spherical trigonometry

The formulas of spherical trigonometry are extremely useful in geodesy. The surface of the Earth, which in first approximation is flat, is in the second approximation (i.e., in a small, but not *very* small, area) the surface of a sphere. Even for the Earth as a whole, the deviations from spherical shape are only 0.3%.

The celestial sphere again may be treated as a precise sphere, the radius R of which is indefinite; in practical computations we often set $R = 1$.

7.1 Spherical excess

Let us look at the sphere presented in Figure 7.1. Assume that the radius of the sphere is 1. On the front side of the sphere in the figure, the union of triangles $A_1 + A_2 + A_3 + A_4$ is a hemisphere, the surface area of which is 2π. Every triangle is bounded by three great circles. The same great circles form on the back side of the sphere (visible in the figure through the front surface) an antipodal triangle of the same size and shape. In the figure, the antipode of A_1, which we might call A_1', is drawn in a lighter shade. All antipodal triangles together, $A_1' + A_2' + A_3' + A_4'$, form again a hemisphere, the back side of the sphere, the surface area of which is also 2π.

Because the surface area of the whole hemisphere is 2π, the area of the "orange-peel piece" bounded by two great circles is equal to $\frac{\alpha}{\pi} \cdot 2\pi$, in which α

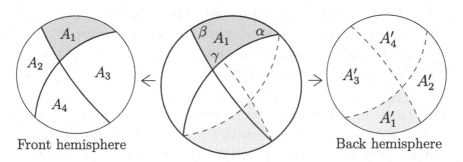

Figure 7.1: Spherical triangles and antipodal triangles on opposite hemispheres.

is the angle between the great circles. We obtain

$$A_1 + A_2 = 2\alpha$$
$$A_1 + A_3 = 2\beta$$
$$A_1 + A_4 = 2\gamma$$

and

$$A_1 + A_2 + A_3 + A_4 = 2\pi.$$

By adding together the first three equations we obtain

$$2A_1 + A_1 + A_2 + A_3 + A_4 = 2\left(\alpha + \beta + \gamma\right),$$

i.e.,

$$A_1 = \alpha + \beta + \gamma - \pi \overset{\text{def}}{=} \varepsilon,$$

in which ε is called the *spherical excess*.

If the radius of the sphere is not 1 but R, we obtain

$$A_1 = \varepsilon R^2 \;\Rightarrow\; \varepsilon = \frac{A_1}{R^2},$$

in which A_1 must be in the same units (e.g., km^2) as R^2. Here, ε is expressed in *radians*. If ε is not already given in radians, we may write

$$\varepsilon\left[\text{unit}\right] = \frac{\rho_{\text{unit}} A_1}{R^2},$$

where ρ_{unit} is the conversion factor of the unit in question, e.g.,

- for degrees: $57.295\,779\,513\,082\,320\,877\,21$, or

- for seconds of arc: $60 \times 60 \times 57.295\,779\,51\ldots$, or

- for gons: $63.661\,977\,236\,758\,134\,308\,01$.

We see that the spherical excess is inversely proportional to the quantity R^2, i.e., *directly proportional to the total curvature of the sphere R^{-2}.* It is also proportional to the surface area of the sphere.

This is a special case of a more general result:

> The change in direction of a vector which is carried in a parallel fashion around the edge of a surface area is, after return to the starting point, the same as the integral of the total curvature over the surface area.

In an equation:

$$\varepsilon = \int_A K \, dS,$$

where S is the surface integration variable and K is the total curvature of the surface named after C. F. Gauss, which thus may vary from place to place. For example, on the surface of an ellipsoid

$$K = \frac{1}{MN}, \tag{7.1}$$

where M is the radius of curvature of the meridian, i.e., the radius of curvature in the South-North direction, and N is the transversal radius of curvature, i.e., the radius of curvature in the East-West direction. Both are dependent on the latitude φ. In a smallish area, the interior geometry of the ellipsoidal surface does not differ much from that of a spherical surface, the radius of which is $R = \sqrt{MN}$.

If a triangle on the surface of a sphere is *small* compared to the radius of the Earth, then also the spherical excess will be small. If $\varepsilon \to 0$, we obtain in the limit $\alpha + \beta + \gamma = \pi$ exactly. We say that a flat surface (or a very small part of a spherical surface) forms a *Euclidean space*, whereas a spherical surface is *non-Euclidean*.

7.2   Surface area of a triangle on a sphere

If the triangle is not very large — i.e., only as large as the triangles of geodetic triangulations generally are, with sides of some 50 km max —, we may calculate its surface area using the formula for a plane triangle:

$$A = \tfrac{1}{2}a \cdot h_a = \tfrac{1}{2}ab \sin \gamma,$$

where h_a is the height of the triangle on the side a, i.e., the straight distance of vertex A from side a.

According to the sine rule for a plane triangle, $b = a \dfrac{\sin \beta}{\sin \alpha}$, and it follows also that

$$A = \frac{a^2 \sin \beta \sin \gamma}{2 \sin \alpha}.$$

Here a is now in metres, and A is thus obtained in square metres. At least for evaluating the spherical excess, these approximate formulas suffice well:

$$\varepsilon = \frac{A}{R^2},$$

in which also an approximate value for R, e.g., $R \approx 6378\,137\,\text{m}$ (the Earth's equatorial radius a_e according to the GRS80 reference ellipsoid), is well sufficient.

7.3   Rectangular spherical triangle

This case is shown in Figure 7.2. A number of simple formulas follow directly from the figure and the separately drawn plane triangles:

$$
\begin{aligned}
EO &= \cos c = \cos a \cos b \\
DE &= \sin c \cos \beta = \sin a \cos b \\
AD &= \sin c \sin \beta = \sin b
\end{aligned}
\tag{7.2}
$$

By interchanging the roles of a and b (and thus of α and β) we obtain furthermore

$$
\begin{aligned}
\sin c \cos \alpha &= \sin b \cos a \\
\sin c \sin \alpha &= \sin a
\end{aligned}
\tag{7.3}
$$

where the first equation yields

$$
\cos \alpha = \cos a \frac{\sin b}{\sin c} = \cos a \sin \beta,
$$

according to the last equation in the set 7.2.

By dividing the first equation in set 7.3 by the second one, we obtain

$$
\cot \alpha = \cot a \sin b
$$

and, similarly, the second equation in set 7.2 by the third one,

$$
\cot \beta = \cot b \sin a.
$$

7.4   General spherical triangle

The equations for the general spherical triangle are obtained by dividing the triangle into two rectangular ones, see Figure 7.3. Here, the third side is $b = b_1 + b_2$.

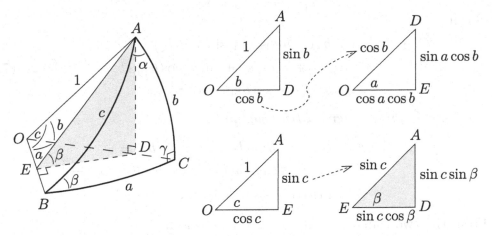

Figure 7.2: A rectangular spherical triangle $\triangle ABC$. Sub-triangles have been exploded into view.

When we apply to the sub-triangles $\triangle CBB'$ and ABB' the equations derived earlier, we obtain

$$\cos a = \cos h \cos b_2,$$
$$\sin a \cos \gamma = \cos h \sin b_2,$$
$$\sin a \sin \gamma = \sin h,$$
$$\cos c = \cos h \cos b_1,$$
$$\sin c \cos \alpha = \cos h \sin b_1,$$
$$\sin c \sin \alpha = \sin h.$$

(7.4)

By substituting

$$\sin b_1 = \sin(b - b_2) = \sin b \cos b_2 - \cos b \sin b_2,$$
$$\cos b_1 = \cos(b - b_2) = \cos b \cos b_2 + \sin b \sin b_2$$

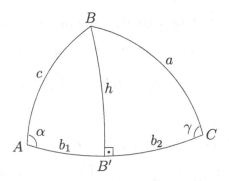

Figure 7.3: A general spherical triangle.

we obtain

$$\cos c = \cos h \left(\cos b \cos b_2 + \sin b \sin b_2\right) =$$
$$= \cos b \left(\cos h \cos b_2\right) + \sin b \left(\cos h \sin b_2\right) = \qquad (7.5)$$
$$= \cos b \cos a + \sin b \sin a \cos \gamma,$$

the *cosine rule for a spherical triangle*, and

$$\sin c \cos \alpha = \cos h \left(\sin b \cos b_2 - \cos b \sin b_2\right) =$$
$$= \sin b \left(\cos h \cos b_2\right) - \cos b \left(\cos h \sin b_2\right) =$$
$$= \sin b \cos a - \cos b \sin a \cos \gamma.$$

From the two "$\sin h$" equations in 7.4 we obtain

$$\sin c \sin \alpha = \sin a \sin \gamma,$$

or more generally

$$\frac{\sin a}{\sin \alpha} = \frac{\sin b}{\sin \beta} = \frac{\sin c}{\sin \gamma},$$

the *sine rule for a spherical triangle*.

For comparison, the corresponding equations for a plane triangle are:

$$c^2 = a^2 + b^2 - 2ab \cos \gamma$$

and

$$\frac{a}{\sin \alpha} = \frac{b}{\sin \beta} = \frac{c}{\sin \gamma}.$$

At least in the case of the sine rule it is clear that, in the limit for a small triangle, $\sin a \mapsto a$, etc., in other words, the spherical sine rule becomes the plane-triangle sine rule. In the case of the cosine rule, showing this is possible by using the approximation for small angles $\cos(a) \approx 1 - \frac{1}{2}a^2$, etc.

7.5   Deriving the equations using vectors in space

If, on a *sphere*, we study a triangle consisting of two points

$$A = \left(\varphi_A = \frac{\pi}{2} - b, \lambda_A = 0\right) \text{ and } B = \left(\varphi_B = \frac{\pi}{2} - a, \lambda_B = \gamma\right)$$

and a *pole* $C = \left(\varphi_C = \frac{\pi}{2}, \lambda_C = \text{arbitrary}\right)$, see Figure 7.4, we may write two three-dimensional vectors (as abstract co-ordinate vectors):

$$\mathbf{x}_A = \begin{bmatrix} x_A \\ y_A \\ z_A \end{bmatrix} = \begin{bmatrix} \cos \varphi_A \\ 0 \\ \sin \varphi_A \end{bmatrix}, \ \mathbf{x}_B = \begin{bmatrix} x_B \\ y_B \\ z_B \end{bmatrix} = \begin{bmatrix} \cos \varphi_B \cos \lambda_B \\ \cos \varphi_B \sin \lambda_B \\ \sin \varphi_B \end{bmatrix}.$$

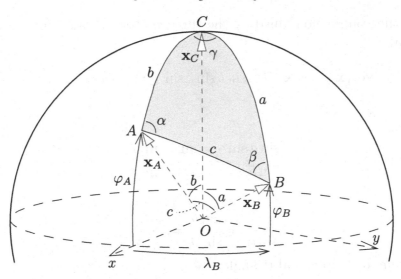

Figure 7.4: A spherical triangle ABC, for deriving the cosine and sine rules by using vectors in space.

The scalar or *dot product* of vectors is defined by the equation

$$\cos c = \langle \mathbf{x}_A \cdot \mathbf{x}_B \rangle = \cos \varphi_A \cos \varphi_B \cos \lambda_B + \sin \varphi_A \sin \varphi_B =$$
$$= \sin b \sin a \cos \gamma + \cos b \cos a,$$

which is the *cosine rule* for a spherical triangle.

The *cross-product* of the vectors, itself a vector, is again

$$\langle \mathbf{x}_A \times \mathbf{x}_B \rangle = \begin{bmatrix} -\sin \varphi_A \cos \varphi_B \sin \lambda_B \\ \sin \varphi_A \cos \varphi_B \cos \lambda_B - \cos \varphi_A \sin \varphi_B \\ \cos \varphi_A \cos \varphi_B \sin \lambda_B \end{bmatrix} =$$
$$= \begin{bmatrix} -\cos b \sin a \sin \gamma \\ \cos b \sin a \cos \gamma - \sin b \cos a \\ \sin b \sin a \sin \gamma \end{bmatrix}.$$

When the third vector is

$$\mathbf{x}_C = \begin{bmatrix} 0 \\ 0 \\ 1 \end{bmatrix},$$

we obtain the *volume* of the parallellepiped spanned by the three vectors — i.e., twice the volume of the tetrahedron $ABCO$ — as follows:

$$\mathrm{Vol}\{\mathbf{x}_A, \mathbf{x}_B, \mathbf{x}_C\} = \langle \langle \mathbf{x}_A \times \mathbf{x}_B \rangle \cdot \mathbf{x}_C \rangle = \sin b \sin a \sin \gamma.$$

However, the volume contained by the three vectors does not depend on their order, so also

$$\text{Vol}\{\mathbf{x}_A, \mathbf{x}_B, \mathbf{x}_C\} = \sin b \sin c \sin \alpha = \sin a \sin c \sin \beta.$$

Division yields

$$\sin a \sin \gamma = \sin c \sin \alpha,$$
$$\sin b \sin \gamma = \sin c \sin \beta,$$

or

$$\frac{\sin a}{\sin \alpha} = \frac{\sin b}{\sin \beta} = \frac{\sin c}{\sin \gamma},$$

the *sine rule* for a spherical triangle.

7.6 Polarization

For every vertex or corner point of a triangle we may define an equator, i.e., a great circle that has as one pole precisely this vertex. If we do so, we obtain three new equators, which themselves again form a triangle. This method is called *polarization* of the triangle, see Figure 7.5.

Because the angular distance between two corner points on the surface of a sphere is a side length, the angle between the planes of two such great circles will be the same as this length. The angle of the polarization triangle is 180° (in radians, π) minus the magnitude of this angle, i.e, the *supplement* of the angle. For example, the length of the side $B'C'$ of the polarization triangle is $\pi - \alpha$, the supplement of the angle $\angle BAC = \alpha$ in the original triangle.

The polarization method is *symmetric*: the original triangle is also the polarization triangle's polarization. The intersection point of two sides of the polarization triangle is at a distance of 90° from both poles, i.e., corners of the original triangle, and the side connecting them is the equator of the intersection point.

For reasons of symmetry, also the side length of a polarization triangle is again the supplement of the original triangle's corresponding vertex angle, i.e., its difference from 180°.

For an arbitrary angle α, it holds that

$$\sin(180° - \alpha) = \sin \alpha$$
$$\cos(180° - \alpha) = -\cos \alpha$$

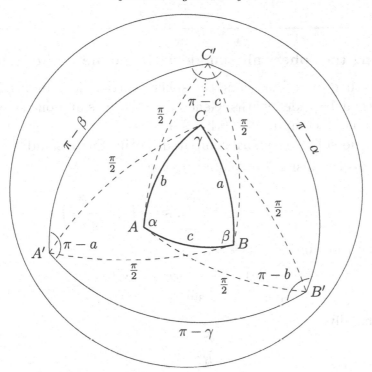

Figure 7.5: Polarization of a spherical triangle. Note that the situation is fully symmetric: if the side $c = AB$ is the equator, then point C' is the corresponding pole, because $AC' = BC' = \frac{\pi}{2}$. And if the side $A'B'$ of the polarized triangle is the equator, then point C of the original triangle is its pole. Also the angles $\angle AC'B'$, $\angle BC'A'$i, etc. are, as angles between an equator and a meridian, all right angles. From this follows, as stated in the text, that the angles of the polarized triangle are the supplements of the corresponding sides of the original triangle, and its sides are the supplements of the angles of the original triangle.

Therefore, one obtains from the spherical cosine rule 7.5 on page 82 the following *polarized version*:

$$-\cos\gamma = (-\cos\beta)(-\cos\alpha) + \sin\beta\sin\alpha(-\cos c)$$

or

$$\cos\gamma = -\cos\beta\cos\alpha + \sin\beta\sin\alpha\cos c,$$

an equation with which we can compute an angle, if two other angles and the length of the side between them are given.

7.7   Solving the spherical triangle with the method of additaments

In the method of additaments, the spherical triangle is reduced to a plane triangle by changing side lengths. Generally, all angles and one side are known. The task is now to compute the other sides.

Because the sides are short compared to the Earth's radius R, we may write the series expansion (remember that $\psi = \dfrac{s}{R}$):

$$\sin \psi = \psi - \tfrac{1}{6}\psi^3 + \ldots \approx \frac{s}{R}\left(1 - \frac{s^2}{6R^2}\right).$$

Now the spherical sine rule is ($s = a, b, c$):

$$\frac{a\left(1 - \partial a\right)}{\sin \alpha} = \frac{b\left(1 - \partial b\right)}{\sin \beta} = \frac{c\left(1 - \partial c\right)}{\sin \gamma},$$

in which generally

$$\partial s = \frac{s^2}{6R^2}, \quad s = a, b, c,$$

i.e.,

$$\partial a = \frac{a^2}{6R^2}, \qquad \partial b = \frac{b^2}{6R^2}, \qquad \partial c = \frac{c^2}{6R^2}.$$

The method works now in this way, that

1. From the known side we subtract its additament ∂s

2. The other sides are computed using the plane-triangle sine rule and known angle values

3. To the sides computed are added *their* additaments.

The additaments are computed from the best available approximate values. If they are initially of poor quality, one improves them by iteration.

The additament correction

$$s' = s\left(1 - \partial s\right)$$

can be modified from a combination of multiplication and subtraction operations to a *simple subtraction* by taking logarithms:

$$\ln s' = \ln s + \ln\left(1 - \partial s\right) = \ln s + \left(0 - \partial s\right) = \ln s - \partial s,$$

i.e., in base-ten algorithms

$$\log_{10} s' = \log_{10} s - M \cdot \partial s = \log_{10} s - \left(\frac{M}{6R^2}\right)s^2,$$

in which $M = \log_{10} e = 0.434\,294\,48$. In the era of logarithm tables, this essentially facilitated the practical computational work.

7.8 Solving the spherical triangle using Legendre's method

In Legendre[1]'s method, the reduction from a spherical triangle to a plane one is done by changing the *angles*. When again all angles and one side are given, we apply the following formula:

$$\frac{a}{\sin\left(\alpha - \varepsilon/3\right)} = \frac{b}{\sin\left(\beta - \varepsilon/3\right)} = \frac{c}{\sin\left(\gamma - \varepsilon/3\right)},$$

i.e., from every angle we subtract *one third of the spherical excess* ε.

It is however important to understand, that the follow-up calculations must always be done *using the original angles* α, β, γ! Removing the spherical excess *only* helps for calculating the unknown sides of the triangle.

Nowadays, these approximative methods — the additaments and Legendre methods — are no longer used, as it is easy to do the computation directly on the computer by the spherical sine rule.

7.9 The half-angle cosine rule

The above cosine rule for spherical triangles (Equation 7.5 on page 82):

$$\cos a = \cos b \cos c + \sin b \sin c \cos \alpha$$

is numerically poorly behaved if the triangle is very small compared to the sphere, i.e., if a, b, c are small.

The triangle Helsinki–Tampere–Turku, for example, is very small compared to the whole Earth, about $\dfrac{200\,\text{km}}{6378\,\text{km}} \approx 0.03$. Then $\cos b \cos c \approx 0.999$, but $\sin b \sin c \approx 0.0009$! We are summing together two terms, one of which is close to 1 and the other about a thousandth part of this, and then we take the result, which will be very close to unity, as the cosine of a to solve a from. This will lead to a serious loss of numerical precision, as the function $\cos a$ is *stationary* around $a = 0$.

For fixing this problem, let us remark first that

$$\cos \alpha = 1 - 2 \sin^2 \frac{\alpha}{2},$$

$$\cos a = 1 - 2 \sin^2 \frac{a}{2},$$

---

[1]Adrien-Marie Legendre (1752–1833) was a French mathematician known, e.g., for Legendre polynomials and functions, which appear in potential theory, e.g., in the study of the Earth's gravity field. He also invented, independently from Gauss, the method of least squares.

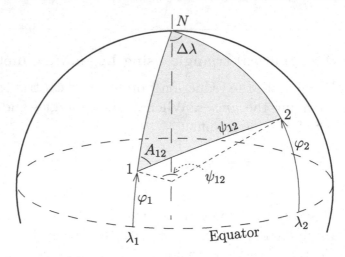

Figure 7.6: Triangle $1 - 2 - N$ on the globe. N is the North Pole.

and

$$\cos b \cos c + \sin b \sin c \cos \alpha = (\cos b \cos c + \sin b \sin c) + \sin b \sin c \, (\cos \alpha - 1) =$$
$$= (\cos b \cos c + \sin b \sin c) - 2 \sin b \sin c \sin^2 \frac{\alpha}{2},$$
$$\cos b \cos c + \sin b \sin c = \cos(b - c) = 1 - 2 \sin^2 \frac{b - c}{2}.$$

After rearranging terms, we obtain an also for small triangles well-behaved *half-angle cosine rule for a spherical triangle* — which, conveniently, only contains sine terms!

$$\sin^2 \frac{a}{2} = \sin^2 \frac{b - c}{2} + \sin b \sin c \sin^2 \frac{\alpha}{2}.$$

7.10   The geodetic forward problem on the sphere

The *geodetic forward problem* is determining the location of a point, when its distance and direction from a point of known location is given.

Applying the cosine and sine rules of a spherical triangle yields

$$\sin \varphi_2 = \sin \varphi_1 \cos \psi_{12} + \cos \varphi_1 \sin \psi_{12} \cos A_{12}$$

and

$$\frac{\sin(\lambda_2 - \lambda_1)}{\sin \psi_{12}} = \frac{\sin A_{12}}{\cos \varphi_2} \implies \lambda_2 = \lambda_1 + \arcsin\left(\frac{\sin \psi_{12} \sin A_{12}}{\cos \varphi_2}\right).$$

Note that here, ψ_{12} is the distance between points 1 and 2 in angular measure as shown in Figure 7.6, or, equivalently, as linear distance on the unit sphere. The linear distance s_{12} over the surface of the real Earth is then $s_{12} = R \cdot \psi_{12}$.

7.11 The geodetic inverse problem on the sphere

The geodetic inverse problem is determining the distance and direction between two points if the locations of both points are given.

By applying the spherical cosine and sine rules , one obtains

$$\cos \psi_{12} = \sin \varphi_1 \sin \varphi_2 + \cos \varphi_1 \cos \varphi_2 \cos (\lambda_2 - \lambda_1) \,,$$
$$\sin A_{12} = \cos \varphi_2 \frac{\sin (\lambda_2 - \lambda_1)}{\sin \psi_{12}} \,. \tag{7.6}$$

Exercises

Assume a *spherical Earth*. The radius of the Earth, in spherical approximation, is $R = a_e = 6378\,137\,\text{m}$ (the semi-major axis of the Earth ellipsoid).

Exercise 7−1: Intersection

Given are the angles α, β and the side between them, c; calculate all the other elements of the triangle (so-called *intersection*).

$$\alpha = 61°45'\,35''.1,$$
$$\beta = 55°15'\,55''.5,$$
$$c = 45\,378.224\,\text{m}.$$

Exercise 7−2: Geodetic forward problem

Given starting point ϕ_1, λ_1, starting azimuth α_{12}, and distance s_{12} (in metres). Calculate the end point ϕ_2, λ_2. This is called the *geodetic forward problem*.

$$\phi_1 = 60°10'\,\text{N},$$
$$\lambda_1 = 24°58'\,\text{E},$$
$$\alpha_{12} = 250^{\text{g}}.0000,$$
$$s_{12} = 1000\,\text{km} = 1000\,000\,\text{m}.$$

Exercise 7 – 3: Geodetic inverse problem

Calculate direction and distance to Mecca, spherical geometry.

Helsinki	$\varphi_1 = 60°10'\,\text{N},$	$\lambda_1 = 24°58'\,\text{E},$
Mecca	$\varphi_2 = 21°27'\,\text{N},$	$\lambda_2 = 39°\,49'\,\text{E}.$

Exercise 7 – 4: Spherical excess (1)

Calculate, for the triangle in the first exercise, the *spherical excess*. Use the angles given and the one you calculated.

Exercise 7 – 5: Spherical excess (2)

Calculate the spherical excess when for the radius of the Earth we assume

$$R = 6356\,752.314\,\text{m},$$

i.e., the semi-minor axis of the Earth ellipsoid. (For this, you have to redo the calculation in the first exercise with the different R value.)

Exercise 7 – 6: Spherical excess (3)

Compare the results of the previous two exercises. What conclusion can we draw?

Chapter 8

The geometry of the ellipsoid of revolution

Figure 8.1 shows the co-ordinates most commonly used together with the ellipsoid of revolution used as a reference ellipsoid: geocentric rectangular co-ordinates X, Y, Z, and geodetic (or geographical) co-ordinates φ, λ, h, in which the geodetic latitude and longitude φ and λ describe, in the point of consideration P, the direction in space of the ellipsoidal normal, and the ellipsoidal height h is the distance of the point of consideration from the surface of the ellipsoid. Between the co-ordinates thus defined, and rectangular co-ordinates, there is a simple mathematical connection.

The geometry of the ellipsoid of revolution is not however, regrettably, quite as simple as that of a sphere. Nevertheless, thanks to rotational symmetry, at least one beautiful invariant can be found.

Solutions to both the geodetic forward problem and the geodetic inverse problem have traditionally been based on series expansions with many terms,

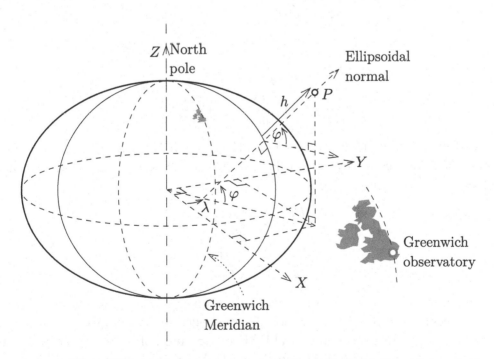

Figure 8.1: The ellipsoid, geocentric co-ordinates X, Y, Z and geodetic co-ordinates φ, λ, h.

the coefficients of which consist also themselves of many terms. In this book, however, we rather use numerical methods, which are conceptually clearer and easier to implement in an error-free fashion.

8.1   The geodesic as a solution of a system of differential equations

We may write, in a small right-angled triangle (dy, dx metric East and North displacements, see Figure 8.2):

$$dx = M(\varphi)\, d\varphi = \cos A\, ds,$$
$$dy = p(\varphi)\, d\lambda = \sin A\, ds,$$
$$dA = \frac{dy}{N \cot \varphi} = \frac{N \cos \varphi\, d\lambda}{N \cot \varphi} = \sin \varphi\, d\lambda,$$

in which $p = N \cos \varphi$ is the distance from the axis of rotation. M is the *meridional radius of curvature*, i.e., the radius of curvature in the North-South direction, and N on the *transversal radius of curvature*, i.e., the radius of curvature in the East-West direction.

The system of equations, normalized:

$$\begin{aligned}
\frac{d\varphi}{ds} &= \frac{\cos A}{M(\varphi)}, \\
\frac{d\lambda}{ds} &= \frac{\sin A}{p(\varphi)}, \\
\frac{dA}{ds} &= \sin \varphi \frac{d\lambda}{ds} = \frac{\sin \varphi \sin A}{p(\varphi)}.
\end{aligned} \tag{8.1}$$

The system 8.1 is valid more generally than on the ellipsoid of revolution: it applies to all bodies of revolution. For a rotationally symmetric body, we have $p(\varphi) = N(\varphi) \cos \varphi$, i.e.,

$$\begin{aligned}
\frac{d\lambda}{ds} &= \frac{\sin A}{N(\varphi) \cos \varphi}, \\
\frac{dA}{ds} &= \frac{\tan \varphi \sin A}{N(\varphi)}.
\end{aligned}$$

If the initial condition given consists of values for $\varphi_1, \lambda_1, A_{12}$, we may obtain the geodesic $\varphi(s), \lambda(s), A(s)$ as a solution parametrized by arc length s. Numerically computing this solution using MATLAB or `octave` software is also relatively straightforward using the ODE[1] (Ordinary Differential Equation) routines.

---

[1] In the `octave` software, the name is `lsode`. Alternatively, one may load the add-on package OdePkg.

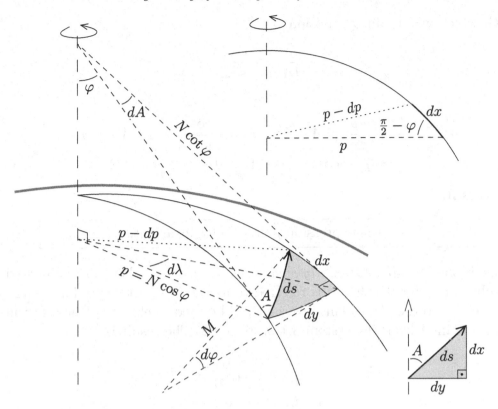

Figure 8.2: The geometry of integrating the geodesic.

Transformation to rectangular form is easy:

$$X(s) = N(\varphi(s)) \cos \varphi(s) \cos \lambda(s),$$
$$Y(s) = N(\varphi(s)) \cos \varphi(s) \sin \lambda(s),$$
$$Z(s) = \frac{b^2}{a^2} N(\varphi) \sin \varphi(s).$$

8.2 An interesting invariant

Let us look more closely at the quantity $p(\varphi) = N(\varphi) \cos \varphi$, the distance of a point from the axis of rotation. Calculate the derivative with respect to s:

$$\frac{dp}{ds} = \frac{dp}{dx} \frac{dx}{ds} = -\sin \varphi \cos A.$$

Now, Equation 8.1 yields

$$\frac{dA}{ds} = \frac{\sin \varphi \sin A}{p},$$

with which we obtain by division

$$\frac{dp}{dA} = -\frac{\cos A}{\sin A} p.$$

Now

$$\frac{d\,(p\sin A)}{dA} = \frac{dp}{dA}\sin A + p\cos A = -\frac{\cos A}{\sin A} p \cdot \sin A + p\cos A =$$
$$= p\,(-\cos A + \cos A) = 0.$$

Result[2]:

> the expression $p\sin A$ is an invariant.

This applies to *all rotationally symmetric bodies*, i.e., also to the ellipsoid of revolution — for which this is referred to as the Clairaut[3] equation — and of course also to the plane. This invariant can be used to eliminate the differential equation in A from the system 8.1 on page 92. The result is

$$\frac{d\varphi}{ds} = \frac{\cos A(\varphi)}{M(\varphi)},$$
$$\frac{d\lambda}{ds} = \frac{\sin A(\varphi)}{p(\varphi)}, \tag{8.2}$$

in which $A(\varphi)$ is computed from the invariant equation

$$\sin A(\varphi) = \sin A_{12}\frac{p(\varphi_1)}{p(\varphi)}. \tag{8.3}$$

The figure of the body is defined by giving the function $p(\varphi)$, or perhaps more practically, $p(Z)$, from which both φ and p are computable.

8.3   The geodetic forward problem

Solving the geodetic forward problem is now done simply by also substituting the given arc length s_{12} into the solution presented above.

---

[2]This invariant may also be interpreted as *conservation of angular momentum*: if a unit-mass billiard ball rolls at a constant speed along a geodesic, its angular momentum around the Z axis, $\langle \mathbf{x} \times \mathbf{v} \rangle_3$, is constant, because, due to the symmetry of the geometry, no torque is acting on the ball around this axis. In this, $\mathbf{x}$ is the vector of place, and $\mathbf{v}$ the velocity vector.

[3]Alexis Claude Clairaut (1713–1765) was a French mathematician, astronomer and geophysicist.

Computing the solution in practice is done most easily by numerical integration. The methods are found in numerical-analysis textbooks and the routines needed in many numerical software libraries.

The classical alternative to numerical methods is offered by the series expansions found in older textbooks. This solution method is in theory more efficient, but complicated.

For the case of the ellipsoid, we may apply Equations 8.2 and 8.3 with the aid of the following equations:

$$M(\varphi) = a \left(1 - e^2\right) \left(1 - e^2 \sin^2 \varphi\right)^{-\frac{3}{2}},$$

$$p(\varphi) = N(\varphi) \cos \varphi = a \left(1 - e^2 \sin^2 \varphi\right)^{-\frac{1}{2}} \cdot \cos \varphi,$$

see Appendix C on page 249.

8.4   The geodetic inverse problem

The straightforward numerical method for solving the reverse problem is *iteration*. Let (φ_1, λ_1) and (φ_2, λ_2) be given. First compute the approximate values[4] $\left(A_{12}^{(1)}, s_{12}^{(1)}\right)$, and solve the geodetic forward problem computing $\left(\varphi_2^{(1)}, \lambda_2^{(1)}\right)$. Thus we obtain *closing errors*

$$\Delta\varphi_2^{(1)} \overset{\text{def}}{=} \varphi_2^{(1)} - \varphi_2,$$

$$\Delta\lambda_2^{(1)} \overset{\text{def}}{=} \lambda_2^{(1)} - \lambda_2.$$

Now these closing errors should be used to compute improved values $\left(A_{12}^{(2)}, s_{12}^{(2)}\right)$, and so on.

The linear dependence between small changes in (A_{12}, s_{12}) and in (φ_2, λ_2) can be approximated using *spherical geometry*.

For this, we need Equations 7.6 on page 89:

$$\cos \psi_{12} = \sin \varphi_1 \sin \varphi_2 + \cos \varphi_1 \cos \varphi_2 \cos(\lambda_2 - \lambda_1),$$

$$\sin \psi_{12} \sin A_{12} = \cos \varphi_2 \sin(\lambda_2 - \lambda_1).$$

Differentiation yields

$$- \sin \psi_{12} \, \Delta\psi_{12} = \left(\sin \varphi_1 \cos \varphi_2 - \cos \varphi_1 \sin \varphi_2 \cos(\lambda_2 - \lambda_1)\right) \Delta\varphi_2 -$$

$$- \cos \varphi_1 \cos \varphi_2 \sin(\lambda_2 - \lambda_1) \, \Delta\lambda_2,$$

$$\sin \psi_{12} \cos A_{12} \, \Delta A_{12} + \cos \psi_{12} \sin A_{12} \, \Delta\psi_{12} =$$

$$= - \sin \varphi_2 \sin(\lambda_2 - \lambda_1) \, \Delta\varphi_2 + \cos \varphi_2 \cos(\lambda_2 - \lambda_1) \, \Delta\lambda_2.$$

---

[4]Approximate values may be computed, e.g., using spherical trigonometry.

So, if we write the matrices

$$\mathbf{A} = \left[\begin{array}{c|c} \begin{array}{c} \sin\varphi_1\cos\varphi_2 - \\ -\cos\varphi_1\sin\varphi_2\cos(\lambda_2-\lambda_1) \\ \hline -\sin\varphi_2\sin(\lambda_2-\lambda_1) \end{array} & \begin{array}{c} -\cos\varphi_1\cos\varphi_2\sin(\lambda_2-\lambda_1) \\ \\ \cos\varphi_2\cos(\lambda_2-\lambda_1) \end{array} \end{array} \right],$$

$$\mathbf{B} = \left[\begin{array}{c|c} -\sin\psi_{12} & 0 \\ \hline \cos\psi_{12}\sin A_{12} & \sin\psi_{12}\cos A_{12} \end{array} \right],$$

we may summarize this as

$$\mathbf{B} \left[\begin{array}{c} \Delta\psi_{12} \\ \Delta A_{12} \end{array} \right] = \mathbf{A} \left[\begin{array}{c} \Delta\varphi_2 \\ \Delta\lambda_2 \end{array} \right].$$

Now we obtain as *iteration equations*:

$$\left[\begin{array}{c} s_{12}^{(i+1)} \\ A_{12}^{(i+1)} \end{array} \right] = \left[\begin{array}{c} s_{12}^{(i)} \\ A_{12}^{(i)} \end{array} \right] + \left[\begin{array}{c} R\Delta\psi_{12}^{(i)} \\ \Delta A_{12}^{(i)} \end{array} \right] = \left[\begin{array}{c} s_{12}^{(i)} \\ A_{12}^{(i)} \end{array} \right] + \left[\begin{array}{cc} R & 0 \\ 0 & 1 \end{array} \right] \mathbf{B}^{-1}\mathbf{A} \left[\begin{array}{c} \Delta\varphi_2^{(i)} \\ \Delta\lambda_2^{(i)} \end{array} \right],$$

in which R is the Earth's mean radius.

With the new values $\left(s_{12}^{(i+1)}, A_{12}^{(i+1)}\right)$ we repeat the computation of the forward geodetic problem to obtain new values $\left(\varphi_2^{(i+1)}, \lambda_2^{(i+1)}\right)$, until convergence. The matrices $\mathbf{A}, \mathbf{B}$ can be computed as needed using the improved approximate values.

8.5   Representations of sphere and ellipsoid

An implicit representation of the *circle* is

$$x^2 + y^2 - a^2 = 0,$$

in which a is the radius (Pythagoras). A parametric representation is

$$x = a\cos\beta,$$
$$y = a\sin\beta.$$

From this we obtain an ellipse by squeezing in the direction of the y axis by a factor $\frac{b}{a}$, i.e.,

$$x = a\cos\beta,$$
$$y = b\sin\beta,$$

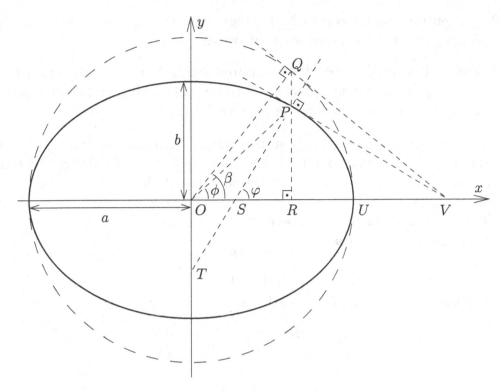

Figure 8.3: The meridian ellipse and different latitude types.

from which we again see that

$$\left(\frac{x}{a}\right)^2 + \left(\frac{y}{b}\right)^2 = \sin^2\beta + \cos^2\beta = 1,$$

i.e.,

$$\frac{x^2}{a^2} + \frac{y^2}{b^2} - 1 = 0$$

is an ellipse, represented implicitly.

8.6 Different types of latitude and relations between them

The latitude on the ellipsoid of revolution may be defined in at least three different ways. Let us look at a cross-section of the ellipsoid of revolution, which itself is an ellipse, the so-called *meridian ellipse*.

In Figure 8.3 we see the following three concepts of latitude:

1. geodetic latitude φ, also geographical latitude: the direction of the ellipsoidal normal with respect to the equatorial plane

2. geocentric latitude ϕ (or ψ): the angle with the equatorial plane of the line connecting the point with the origin

3. reduced latitude β: point P is shifted straight in the y direction to a point Q on the circle surrounding the meridian ellipse. The geocentric latitude of point Q is now the reduced latitude of point P.

The reduced latitude is used only in theoretical contexts. On maps, geodetic (i.e., geographical or ellipsoidal) latitudes are used. The geocentric latitude again is used in practice only in connection with satellite and space geodesy.

The *longitude* λ is the same, no matter whether we are talking about geodetic, geocentric or reduced co-ordinates.

In Figure 8.3 we see that

$$\frac{PR}{QR} = \frac{b}{a}.$$

In the triangles ORQ and ORP we have

$$\tan\beta = \frac{QR}{OR} \quad \text{and} \quad \tan\phi = \frac{PR}{OR},$$

i.e.,

$$\frac{\tan\phi}{\tan\beta} = \frac{PR}{QR} = \frac{b}{a},$$

or

$$\tan\beta = \frac{a}{b}\tan\phi.$$

In the triangles RVQ, RVP:

$$\tan\left(\frac{\pi}{2} - \beta\right) = \frac{QR}{VR} \quad \text{and} \quad \tan\left(\frac{\pi}{2} - \varphi\right) = \frac{PR}{VR},$$

i.e.,

$$\frac{\cot\varphi}{\cot\beta} = \frac{PR}{QR} = \frac{b}{a},$$

or

$$\tan\varphi = \frac{a}{b}\tan\beta.$$

Combining still yields

$$\tan\varphi = \frac{a^2}{b^2}\tan\phi.$$

8.7 Different measures for the flattening

The *flattening of the reference ellipsoid* is described by at least three different geometric quantities: the flattening or flattening ratio f, the first eccentricity e, and the second eccentricity e', the definitions of which are

$$f = \frac{a-b}{a}, \qquad e^2 = \frac{a^2 - b^2}{a^2}, \qquad (e')^2 = \frac{a^2 - b^2}{b^2}. \tag{8.4}$$

These quantities are mathematically equivalent. Between them, the following relationships exist:

$$(e')^2 = \frac{a^2}{b^2} e^2 \iff be' = ae \iff e^2 = \frac{b^2}{a^2} (e')^2, \tag{8.5}$$

as well as

$$1 - e^2 = 1 - \frac{a^2 - b^2}{a^2} = \frac{b^2}{a^2} \quad \text{and} \quad 1 - f = 1 - \frac{a-b}{a} = \frac{b}{a},$$

i.e.,

$$1 - e^2 = (1 - f)^2 \implies f = 1 - \sqrt{1 - e^2}. \tag{8.6}$$

8.8 Co-ordinates on the meridian ellipse

Compute the co-ordinates x, y of point P on the meridian ellipse as follows.

The distance PT is designated with the symbol N, i.e., the *transversal radius of curvature*, the curvature of the ellipsoid surface in the East-West direction.

Now we have

$$x = N \cos \varphi. \tag{8.7}$$

Also

$$PR = OR \tan \phi = OR \frac{b^2}{a^2} \tan \varphi = N \cos \varphi \cdot \left(1 - e^2\right) \tan \varphi$$

using $OR = x = N \cos \varphi$. The end result, because $y = PR$, is

$$y = N \left(1 - e^2\right) \sin \varphi. \tag{8.8}$$

Equations 8.7 and 8.8 represent a description of the meridian ellipse as a function of geodetic latitude φ. *Remember* that N is a function of φ, i.e., *it is not a constant!* In fact

$$\frac{x^2}{a^2} + \frac{y^2}{b^2} = \frac{N^2}{a^2}\cos^2\varphi + \frac{N^2\left(1-e^2\right)^2}{b^2}\sin^2\varphi =$$

$$= \frac{N^2}{a^2}\cos^2\varphi + \frac{N^2}{a^2}\frac{b^2}{a^2}\sin^2\varphi = \frac{N^2}{a^2}\left(\cos^2\varphi + \frac{b^2}{a^2}\sin^2\varphi\right) = 1;$$

from the latter condition follows

$$N(\varphi) = \frac{a}{\sqrt{1 - e^2\sin^2\varphi}},$$

by substituting definition 8.4 of e^2.

8.9   Three-dimensional rectangular co-ordinates on the reference ellipsoid

The equations obtained above are easily generalized: if x and y are co-ordinates in the meridian section, then the three-dimensional rectangular co-ordinates are

$$X = x\cos\lambda = N\cos\varphi\cos\lambda,$$
$$Y = x\sin\lambda = N\cos\varphi\sin\lambda,$$
$$Z = y = N\left(1 - e^2\right)\sin\varphi.$$

If we look at points not on the surface of the reference ellipsoid, but inside or outside it in space, we may write

$$
\begin{aligned}
X &= (N + h)\cos\varphi\cos\lambda, \\
Y &= (N + h)\cos\varphi\sin\lambda, \\
Z &= \left(N\left(1 - e^2\right) + h\right)\sin\varphi.
\end{aligned}
\tag{8.9}
$$

Here h is the straight distance from the ellipsoid surface ("ellipsoidal height"). This quantity is interesting because we may say that satellite positioning equipment is able to measure just this quantity. More precisely, they measure co-ordinates X, Y, Z and compute from these the parameter h (and φ, $N(\varphi)$ and λ), which succeeds if the defining parameters a and e of the reference ellipsoid are known.

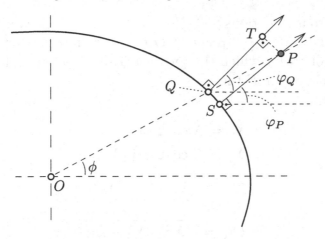

Figure 8.4: From geocentric to geodetic latitude.

8.10 Computing geodetic co-ordinates from rectangular ones

The inverse problem to that described by Equations 8.9 is not quite simple to solve. Closed solutions are known, but they are complicated. Iterative solutions of the problem, based on using Equations 8.9 directly, are fully possible and often done.

Computing latitude λ is extremely simple:

$$\tan \lambda = \frac{Y}{X}.$$

A possible stumbling block is determining the correct quadrant of λ — use the function `atan2` in your programming environment, if it is on offer.

Determining φ and h is more complicated. See Figure 8.4, in which the rectangular co-ordinates X, Y, Z of point P are known and the geodetic co-ordinates φ, λ, h may be computed.

Compute first the *geocentric latitude* by the equation

$$\tan \phi = \frac{Z}{\sqrt{X^2 + Y^2}}$$

and the *geocentric distance* (radius) by the equation

$$OP = \sqrt{X^2 + Y^2 + Z^2}.$$

If the point P *were* to lie on the reference ellipsoid, we *could* calculate its geodetic latitude φ_P by the following equation:

$$\tan \varphi_P = \frac{Z}{(1 - e^2) \sqrt{X^2 + Y^2}}$$

(proof directly from Equations 8.9). Now, if P is above the ellipsoid, we obtain in this way the latitude φ_Q of point Q, Q being the intersection of the ellipsoid and the radius of P, for which the (geometrically obvious) relationship given above is the same as for P there.

Now we compute

$$X_Q = N \cos \varphi_Q \cos \lambda,$$
$$Y_Q = N \cos \varphi_Q \sin \lambda,$$
$$Z_Q = N \left(1 - e^2\right) \sin \varphi_Q,$$

from which

$$OQ = \sqrt{X_Q^2 + Y_Q^2 + Z_Q^2}$$

and thus

$$PQ = OP - OQ.$$

Furthermore, in the little triangle TQP we may calculate

$$\angle TQP = \varphi_Q - \phi$$

and thus

$$TP = PQ \sin \left(\varphi_Q - \phi\right),$$
$$TQ = PQ \cos \left(\varphi_Q - \phi\right).$$

Now

$$h = PS \approx TQ$$

and (exploiting that, in spherical approximation, $\frac{TP}{PO}$ is roughly the angle in radians by which the direction of the ellipsoidal normal changes when travelling from point P to point T):

$$\varphi_P \approx \varphi_Q - \frac{TP}{PO}, \tag{8.10}$$

or perhaps a tiny bit more precisely[5]

$$\varphi_P \approx \varphi_Q - \frac{TP}{PO} \cos(\varphi_Q - \phi). \tag{8.11}$$

The method is in practice fairly accurate. If $h = 8000 \, \text{m}$ and $\varphi = 45°$, then $TP \approx 26 \, \text{m}$, so then between the solutions for φ_P, 8.10 and 8.11, there is a difference of $0.1 \, \text{mm}$[6], which is also the order of magnitude of the error possibly remaining in the different solutions. The approximation $PS \approx TQ$ contains an error of $0.05 \, \text{mm}$[7].

---

[5]Also, instead of PO, one should take $M(\varphi_Q) + h$, in which M is the meridional curvature.
[6]Linearly $TP(1 - \cos(\varphi_Q - \phi))$.
[7]$\approx \frac{1}{2} \frac{TP^2}{OP}$.

8.11   Length of the meridian arc

The length of the meridian arc is a quantity needed, e.g., in the UTM and Gauss–Krüger projections and which is calculated by integration. Earlier we defined the quantity N, the *transversal radius of curvature*. The other radius of curvature of the Earth is the *meridional radius of curvature M*. When it is given as a function of latitude φ, an element of distance ds is calculated with the formula

$$ds = M(\varphi)\, d\varphi.$$

Now we can calculate the length of the meridian arc as follows:

$$s(\varphi_0) = \int_0^{\varphi_0} M(\varphi)\, d\varphi.$$

On the reference ellipsoid

$$M(\varphi) = \frac{a\left(1 - e^2\right)}{\left(1 - e^2 \sin^2 \varphi\right)^{\frac{3}{2}}},$$

i.e.,

$$s\left(\varphi_0\right) = a\left(1 - e^2\right) \int_0^{\varphi_0} \left(1 - e^2 \sin^2 \varphi\right)^{-\frac{3}{2}} d\varphi.$$

Here, the last coefficient may be expanded into a series — because $e^2 \sin^2 \varphi \ll 1$ — and integrated termwise. Of course also numerical computation is possible, and nowadays that may even be the best alternative. MATLAB offers for this purpose the useful `QUAD` (quadrature) routines[8], as does also `octave`.

8.12   The reference ellipsoid and the gravity field

The reference ellipsoid is a purely geometric description of the figure of the Earth. However, a reference *system* comprises also a gravity field for the Earth. The reference ellipsoid is an approximation, not of the body of the solid Earth, but of its mean sea level. And actually not even of the true mean sea level, but of the theoretical mean sea level, which would be realized if sea water was affected only by the gravity field of the Earth, and it would settle into a minimum-energy state of rest. In other words, we are dealing with the *geoid* of the Earth.

---

[8]Soon to be replaced by INTEGRAL.

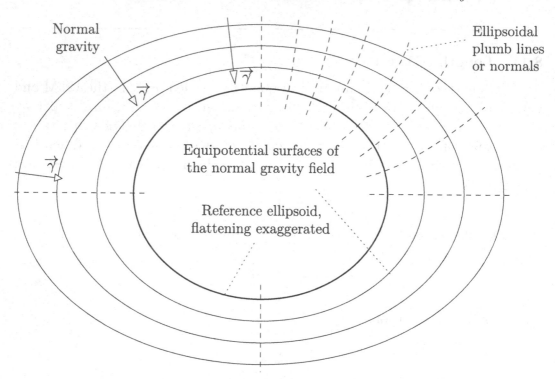

Figure 8.5: The normal gravity field of the Earth. The reference ellipsoid is one of its equipotential surfaces.

Thus, the reference ellipsoid describes the gravity field of the Earth as a fully regular, rotationally symmetric field, one *equipotential surface*, i.e., set of points having the same potential value, of which is the surface of the ellipsoid itself. This model gravity field is called the *normal gravity field*. A cross-section of this normal gravity field, complete with lines of force and equipotential surfaces, is shown in Figure 8.5.

Gravity is a vector, of dimensionality "acceleration" ("the acceleration of free fall"), which is everywhere perpendicular to the equipotential surfaces. Its direction is also called the direction of the *plumb line*. It is the *gradient* of the gravity potential. Note that gravity contains also the centrifugal force caused by the Earth's rotation! We have thus the resultant of attraction (gravitation) and centrifugal force. More abstractly we could say that gravity is gravitation expressed in a system co-rotating with the Earth, an ECEF (Earth centred, Earth-fixed) system. All this applies also for the normal gravity field.

We will present here matters related to the Earth's gravity field only superficially, and restrict ourselves to things that are related to geometry. We follow the presentation used in the textbook by Heiskanen and Moritz (1967, chapter 2), and from that, only some useful formulas. Partly the same math is also found in the book by Torge (2001), subsection 4.2.2.

If the mass enclosed by the ellipsoid GM is known, we may directly calculate some of its gravimetric parameters, e.g., normal gravity, and the accelerations of free fall both at the equator and at the poles (Heiskanen and Moritz, 1967, equations 2-73 and 2-74):

$$\gamma_e = \frac{GM}{ab}\left(1 - m - \frac{m}{6}\frac{e'q_0'}{q_0}\right),$$

$$\gamma_p = \frac{GM}{a^2}\left(1 + \frac{m}{3}\frac{e'q_0'}{q_0}\right).$$

This requires that we know some parameters related to the flattening, like (Heiskanen and Moritz, 1967, equations 2-67 and 2-71):

$$q_0'(e') = 3\left(1 + \frac{1}{e'^2}\right)\left(1 - \frac{1}{e'}\arctan e'\right) - 1$$

and (Heiskanen and Moritz, 1967, equations 2-58 and 2-71):

$$2q_0(e') = \left(1 + \frac{3}{e'^2}\right)\arctan e' - \frac{3}{e'} \tag{8.12}$$

as well as (Heiskanen and Moritz, 1967, equation 2-70):

$$m = \frac{\omega^2 a^2 b}{GM}. \tag{8.13}$$

The definition of the quantity e', known as the *second eccentricity*, is

$$(e')^2 = \frac{a^2 - b^2}{b^2},$$

when the definition of the ordinary (first) eccentricity is

$$e^2 = \frac{a^2 - b^2}{a^2},$$

see Definitions 8.4 on page 99. We also use the identity $be' = ae$, which was earlier derived in Equation 8.5 on page 99.

In a later Section 10.2 on page 128 these equations are used in the iterative calculation of e^2 and all other parameters on the GRS80 reference ellipsoid.

Define now the gravity flattening as follows:

$$f^* \stackrel{\text{def}}{=} \frac{\gamma_p - \gamma_e}{\gamma_e}.$$

Clairaut found in 1738 a beautiful equation between the geometric and gravity flattenings of the Earth. It is often given in the following approximate form:

$$f + f^* \approx \frac{5}{2}m,$$

in which the number m is often also approximated by the equation

$$m \approx \frac{\omega^2 a}{\gamma_e},$$

the ratio between centrifugal force and gravity on the equator (Heiskanen and Moritz, 1967, equations 2-99, 2-100).

Laplace attempted already in 1799 to compute a value for the flattening of the Earth by using gravity values, available already then, measured in different locations in South America and Europe, like in France and also in Lapland (Laplace, 1799, page 147). The measurements had been done using a one-second wire pendulum, and the results of the measurements were stated as lengths of the wires. Laplace reduced the measurements to sea level, and computed for the flattening f an upper bound[9] of $f = \frac{1}{321.48}$ and a most likely value of $f = \frac{1}{335.78}$, and concluded that a homogeneous mass distribution within the Earth, as Newton had assumed, (see Section 1.6.1 on page 10), is not possible.

A simple and elegant formula for normal gravity on the surface of the reference ellipsoid is that of Somigliana[10] and Pizzetti[11]:

$$\gamma = \frac{a\gamma_e \cos^2 \varphi + b\gamma_p \sin^2 \varphi}{\sqrt{a^2 \cos^2 \varphi + b^2 \sin^2 \varphi}}.$$

A similar equation may be written using the *reduced latitude*, see Section 8.6 on page 97:

$$\gamma = \frac{a\gamma_p \sin^2 \beta + b\gamma_e \cos^2 \beta}{\sqrt{a^2 \sin^2 \beta + b^2 \cos^2 \beta}}.$$

The equivalence of these equations is easy to show.

---

[9]Note that also the modern value for the flattening, $f = \dfrac{1}{298.257}$, contradicts this result from Laplace; the difference is however only 7%.

[10]Carlo Somigliana (1860–1955) was an Italian mathematician and geophysicist.

[11]Paolo Pizzetti (1860–1918) was an Italian geodesist, astronomer and geophysicist.

Exercises

Exercise 8 − 1: Meridian arc length

Let there be two points A and B on the meridian $\lambda = 24°$: $\varphi_A = 60°$ and $\varphi_B = 70°$ (geodetic co-ordinates on the GRS80 ellipsoid). The heights of the points from the GRS80 reference ellipsoid are 0.

- **a.** Calculate the parameter e^2 for both the GRS80 and the Hayford ellipsoid. The defining parameters of both are found in Section 10.3.

- **b.** Calculate the radius of curvature on the meridional direction $M(\varphi)$ in the points A, B and the midpoint C $(\varphi_C = 65°, \lambda_C = 24°)$.

- **c.** Calculate the length of the meridian from A to B. Use *Simpson's rule*:

$$\int_{\varphi_A}^{\varphi_B} f(\varphi)\, d\varphi \approx \tfrac{1}{3}\Delta\varphi\big(f(\varphi_A) + 4f(\varphi_C) + f(\varphi_B)\big),$$

  in which $\Delta\varphi = \varphi_C - \varphi_A = \varphi_B - \varphi_C$.

- **d.** Calculate the length of the straight chord in space from A to B. How does the answer differ from the previous case?

Exercise 8 − 2: Gall's projection

James Gall's so-called orthographic cylinder projection was designed using a sphere as the reference surface. Then, the equations for it and many other equivalent cylinder projections[12] are

$$x = R\lambda,$$
$$y = cR\sin\varphi, \tag{8.14}$$

in which c is an arbitrary constant defining the latitude at which the projection is conformal: e.g., $c = 1$ for Lambert's equivalent cylinder projection, $c = 2$ for the Gall projection, etc.

Let us write Equation 8.14 into this form:

$$dy = cR\cos\varphi\, d\varphi = c\, p(\varphi)\, d\varphi. \tag{8.15}$$

In this, $p(\varphi)$ is the distance from the axis of rotation. Equation 8.15 applies generally for all rotationally symmetric bodies.

---

[12]Behrmann equal-area; Smyth equal-surface, alias Craster rectangular equal-area; Trystan Edwards; Hobo-Dyer; Baltasart. A beloved child has many names, and "great minds think alike."

For the ellipsoid of revolution $p(\varphi) = N(\varphi) \cos \varphi$, yielding

$$dy = cN(\varphi) \cos \varphi \, d\varphi = \frac{ca}{1 - e^2 \sin^2 \varphi} \cos \varphi \, d\varphi.$$

Exercise: integrate this equation, i.e., derive the function $y(\varphi)$ for the ellipsoid of revolution.

Exercise 8 – 3: Co-ordinate transformation

Compute the geographical co-ordinates of point A, given on the GRS80 reference ellipsoid in the first exercise — i.e., latitude 60°, longitude 24°, height 0 — also on the Hayford ellipsoid, if additionally it is given that the centre of the Hayford ellipsoid in the GRS80 system is (according to M. Ollikainen)

$$\begin{bmatrix} X_0 \\ Y_0 \\ Z_0 \end{bmatrix} = \begin{bmatrix} -93.477\,\text{m} \\ -103.453\,\text{m} \\ -123.431\,\text{m} \end{bmatrix}$$

(Hint: compute first the XYZ co-ordinates from the geographical co-ordinates given on the GRS80 reference ellipsoid, then refer them to the centre of the Hayford reference ellipsoid; then, using the Hayford parameter values, compute φ, λ, h, see Section 8.10).

Chapter 9

Three-dimensional co-ordinates and transformations

Three-dimensional rectangular co-ordinates are extensively used in modern geodesy, because their computation is, thanks to computer technology, easy. Three-dimensional co-ordinates come in two main types:

1. geocentric co-ordinates X, Y, Z

2. topocentric co-ordinates x, y, z, also called NEU (North, East, Up).

The origin of geocentric co-ordinates is the *centre of mass of the Earth* (or a point very close to it). The origin of topocentric co-ordinates is *one's own location* on the Earth's surface, i.e., the point being observed or in which observations are being made.

Three-dimensional locations may be written as *vectors*, for example

$$\mathbf{X} = \mathbf{X_0} + \mathbf{x},$$

in which $\mathbf{X_0}$ is the topocentric origin, i.e., the geocentric vector describing our own location.

We may expand vectors into components or co-ordinates on a *basis*, for example

$$\mathbf{X} = X\mathbf{i} + Y\mathbf{j} + Z\mathbf{k}.$$

In this notation, the unit vectors $\{\mathbf{i}, \mathbf{j}, \mathbf{k}\}$ form an *orthonormal basis*, which spans Euclidean space $\mathbb{R}^3$ or $\mathbb{E}^3$. Let this basis be geocentric, and let an alternative, topocentric basis be $\{\mathbf{i}', \mathbf{j}', \mathbf{k}'\}$. Then

$$\mathbf{x} = x\mathbf{i}' + y\mathbf{j}' + z\mathbf{k}',$$

in which the orthonormal basis $\{\mathbf{i}', \mathbf{j}', \mathbf{k}'\}$ is in a different orientation.

An alternative but useful representation of vectors is as *column vectors of co-ordinates* on some basis, like

$$\overline{\mathbf{X}} = \begin{bmatrix} X \\ Y \\ Z \end{bmatrix}, \quad \overline{\mathbf{x}} = \begin{bmatrix} x \\ y \\ z \end{bmatrix}.$$

Between different co-ordinate frames there exist *transformations*, which may be represented by size 3×3 matrices. A common transformation is rotation.

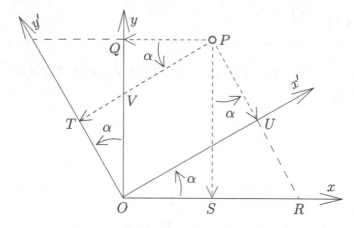

Figure 9.1: Rotation in the plane.

9.1   About rotations and rotation matrices

When we change the orientation of the axes of a co-ordinate system, we have, in rectangular co-ordinates, a *multiplication* of co-ordinate vectors with a *rotation matrix*.

Let us first look at the plane (x, y), see Figure 9.1.

A new co-ordinate is calculated by the formula

$$x'_P = OU = OR \cos \alpha,$$

in which

$$OR = OS + SR = x_P + PS \tan \alpha = x_P + y_P \tan \alpha.$$

Substitution yields

$$x'_P = (x_P + y_P \tan \alpha) \cos \alpha = x_P \cos \alpha + y_P \sin \alpha.$$

In the same way is obtained

$$y'_P = OT = OV \cos \alpha,$$

in which

$$OV = OQ - VQ = y_P - PQ \tan \alpha = y_P - x_P \tan \alpha,$$

from which follows

$$y'_P = (y_P - x_P \tan \alpha) \cos \alpha = -x_P \sin \alpha + y_P \cos \alpha.$$

Summarizingly in a matric equation:

$$\begin{bmatrix} x' \\ y' \end{bmatrix} = \begin{bmatrix} \cos\alpha & \sin\alpha \\ -\sin\alpha & \cos\alpha \end{bmatrix} \begin{bmatrix} x \\ y \end{bmatrix}.$$

The easiest way to ascertain the place of the minus sign is by drawing both pairs of axes on paper, marking the angle α, and inferring graphically whether a point on the positive x axis (i.e., $y = 0$) has a new y' co-ordinate that is positive or negative: in the above case one obtains

$$y' = \begin{bmatrix} -\sin\alpha & \cos\alpha \end{bmatrix} \begin{bmatrix} x \\ 0 \end{bmatrix} = -\sin\alpha \cdot x,$$

i.e., $y' < 0$, so the minus sign really is in the bottom left corner.

9.2 Chaining matrices in three dimensions

In a three-dimensional co-ordinate system, we may write the two-dimensional rotation matrix in the following way:

$$\begin{bmatrix} x' \\ y' \\ z' \end{bmatrix} = \begin{bmatrix} \cos\alpha & \sin\alpha & 0 \\ -\sin\alpha & \cos\alpha & 0 \\ 0 & 0 & 1 \end{bmatrix} \begin{bmatrix} x \\ y \\ z \end{bmatrix}.$$

In other words, the z co-ordinate is copied over as such $z' = z$, while x and y transform amongst themselves by the above equation.

If there are multiple successive co-ordinate transformations, one obtains the final transformation by "chaining" the transformation matrices. If we write

$$\overline{\mathbf{x}}'' = \mathbf{S}\overline{\mathbf{x}}', \ \overline{\mathbf{x}}' = \mathbf{R}\overline{\mathbf{x}},$$

in which

$$\mathbf{R} = \begin{bmatrix} \cos\alpha & \sin\alpha & 0 \\ -\sin\alpha & \cos\alpha & 0 \\ 0 & 0 & 1 \end{bmatrix}, \ \mathbf{S} = \begin{bmatrix} 1 & 0 & 0 \\ 0 & \cos\beta & \sin\beta \\ 0 & -\sin\beta & \cos\beta \end{bmatrix},$$

then (*associativity* of matrix multiplication)

$$\overline{\mathbf{x}}'' = \mathbf{S}\left(\mathbf{R}\overline{\mathbf{x}}\right) = \left(\mathbf{SR}\right)\overline{\mathbf{x}},$$

i.e., the matrices are combined by multiplication.

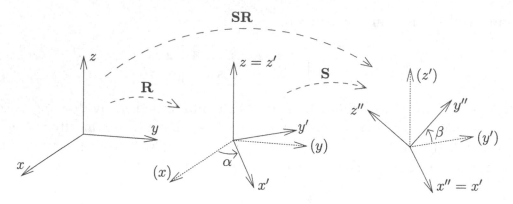

Figure 9.2: Rotation operators acting on the co-ordinate frame are associative, i.e., the law of associativity of rotation matrices applies.

In the sequel, we leave off the overbars when no risk of confusion exists. Remember that

$$\mathbf{RS} \neq \mathbf{SR},$$

i.e., matrices and co-ordinate transformations are not (generally) commutative[1], so there is no *commutativity of matrix multiplication*!

See Figure 9.2.

9.3 Orthogonal matrices

Rotation matrices are *orthogonal*:

$$\mathbf{RR}^\mathsf{T} = \mathbf{R}^\mathsf{T}\mathbf{R} = \mathbf{I}, \tag{9.1}$$

their inverse matrix is the same as their transpose.

As an example

$$\begin{bmatrix} \cos\alpha & \sin\alpha \\ -\sin\alpha & \cos\alpha \end{bmatrix}^{-1} = \begin{bmatrix} \cos\alpha & -\sin\alpha \\ \sin\alpha & \cos\alpha \end{bmatrix} \left(= \begin{bmatrix} \cos(-\alpha) & \sin(-\alpha) \\ -\sin(-\alpha) & \cos(-\alpha) \end{bmatrix} \right),$$

fully understandable, as it is a rotational motion around the same axis. In it, the amount of rotation is the same, but in the *opposite direction*.

Equation 9.1 can be written as follows:

$$\sum_{i=1}^{n} R_{ij} R_{ik} = \begin{cases} 1, & j = k, \\ 0, & j \neq k. \end{cases}$$

---

[1]Two-dimensional rotations *are* in fact commutative, like the addition of rotation angles. This also follows from the fact that they can be described using complex numbers.

The columns of a rotation matrix are *orthonormal,* their norm (length) is 1 and they are mutually orthogonal. This can be seen in the case of the example matrix:

$$\sqrt{\cos^2 \alpha + \sin^2 \alpha} = 1,$$
$$\cos \alpha \cdot \sin \alpha + (-\sin \alpha) \cdot \cos \alpha = 0.$$

One often comes across other orthogonal matrices:

1. the axis mirroring matrix, e.g.:

$$\mathbf{M}_2 \overset{\text{def}}{=} \begin{bmatrix} 1 & 0 & 0 \\ 0 & -1 & 0 \\ 0 & 0 & 1 \end{bmatrix},$$

   which reverses the direction of the y co-ordinate, i.e., its algebraic sign

2. the axes permutation matrix:

$$\mathbf{P}_{12} \overset{\text{def}}{=} \begin{bmatrix} 0 & 1 & 0 \\ 1 & 0 & 0 \\ 0 & 0 & 1 \end{bmatrix}$$

3. the *inversion* of all axes:

$$\mathbf{X} \overset{\text{def}}{=} \begin{bmatrix} -1 & 0 & 0 \\ 0 & -1 & 0 \\ 0 & 0 & -1 \end{bmatrix}.$$

Both $\mathbf{M}$ and $\mathbf{P}$ differ from rotation matrices in the sense that their *determinant* is -1, when for rotation matrices it is $+1$. The determinant of matrix $\mathbf{X}$ is $(-1)^n$, n being the number of dimensions, three in the above case. A determinant of -1 means that the transformation switches a right-handed co-ordinate frame into a left-handed one, and vice versa.

If we multiply, for example, $\mathbf{M}_2$ and $\mathbf{P}_{12}$, we obtain

$$\mathbf{M}_2 \mathbf{P}_{12} = \begin{bmatrix} 0 & 1 & 0 \\ -1 & 0 & 0 \\ 0 & 0 & 1 \end{bmatrix}.$$

The determinant of this is $+1$. So, the above is again a rotation matrix:

$$\mathbf{R}_3 (90°) = \begin{bmatrix} \cos 90° & \sin 90° & 0 \\ -\sin 90° & \cos 90° & 0 \\ 0 & 0 & 1 \end{bmatrix}!$$

> All orthogonal transformations with a positive determinant are rotations.

Without proof we still note that *all* orthogonal transformations can be written either as a rotation around a certain axis, or as a mirroring through a certain plane.

9.4   Topocentric systems

These systems, which are connected to the horizon of the observation site, are sometimes called "local astronomical" systems. Note that, whereas a geocentric system is uniquely defined, i.e., there is only one such system of a certain type, there are as many local systems as there are points on the Earth's surface, i.e., an infinite number of them.

The axes of the system are the following:

1. The z axis points to the local zenith, straight up.

2. The x axis points to the local North.

3. The y axis is perpendicular to the other two, and points East.

We also speak of a NEU ("North-East-Up") system.

In Figure 9.3 is shown the situation of the local topocentric system in the global context.

The spherical co-ordinates of the system are as follows:

o the azimuth A, typically, though not always, counted clockwise from the North

o the zenith angle ζ, alternatively the elevation angle $\eta = 90° - \zeta = 100\,\mathrm{gon} - \zeta$

o the distance or (slant) range s.

The conversion between the topocentric spherical co-ordinates and the rectangular co-ordinates of point P is:

$$
\begin{bmatrix} x \\ y \\ z \end{bmatrix}_{\mathrm{T}} = \begin{bmatrix} s\cos A \sin\zeta \\ s\sin A \sin\zeta \\ s\cos\zeta \end{bmatrix}_{\mathrm{T}}.
$$

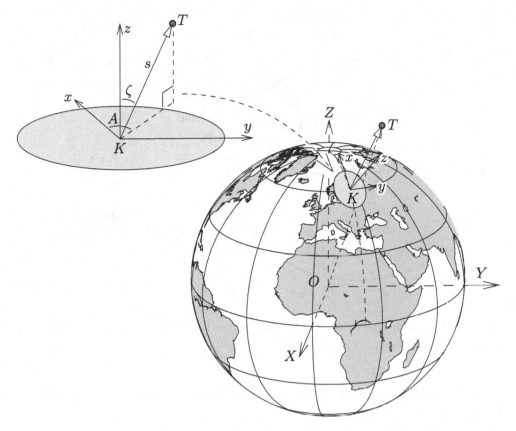

Figure 9.3: Topocentric and geocentric co-ordinates.

The reverse conversion:

$$\zeta = \arctan \frac{\sqrt{x^2 + y^2}}{z},$$

$$A = 2 \arctan \frac{y}{x + \sqrt{x^2 + y^2}}.$$

The latter formula is also known as the *half-angle formula*, and by using it one obtains always the correct value for A. The naive use of the arctan function may easily give an angle value which differs by 180° from the correct value[2]. The result is in the interval $(-180°, 180°]$ and negative values are made positive by adding 360°.

9.4.1 Geocentric to topocentric and back

Let (X, Y, Z) be a geocentric co-ordinate frame and (x, y, z) a topocentric *instrument co-ordinate frame* (so, the x axis points in the zero direction of the

---

[2]Another way to circumvent this problem is to use the function `atan2` available in many programming languages.

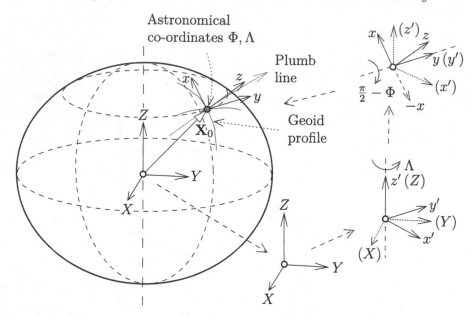

Figure 9.4: From the geocentric system to the topocentric system. The matrix $\mathbf{R}_1$ mentioned in the text has been left out.

instrument, rather than to the North; for a theodolite, this is the zero direction on the horizontal circle.)

In this case, we may symbolically write

$$\overline{\mathbf{x}} = \mathbf{R}_1 \left(\mathbf{M}_1 \mathbf{R}_2 \right) \mathbf{R}_3 \left(\overline{\mathbf{X}} - \overline{\mathbf{X}}_0 \right),$$

in which the rotation matrices $\mathbf{R}_3$, $\mathbf{R}_2$, $\mathbf{R}_1$ are operating in succession to transform $\overline{\mathbf{X}}$ into $\overline{\mathbf{x}}$. See Figure 9.4. Here, $\overline{\mathbf{X}}_0$ contains the co-ordinates of the local origin expressed in the geocentric system. The mirroring matrix $\mathbf{M}_1$ is needed because the topocentric system is left-handed, whereas the geocentric system is right-handed. Its definition is (mirroring of the x axis)

$$\mathbf{M}_1 = \begin{bmatrix} -1 & 0 & 0 \\ 0 & 1 & 0 \\ 0 & 0 & 1 \end{bmatrix}.$$

Considering all this, the reverse transformation chain is

$$\overline{\mathbf{X}} = \overline{\mathbf{X}}_0 + \mathbf{R}_3^\mathsf{T} \left(\mathbf{R}_2^\mathsf{T} \mathbf{M}_1 \right) \mathbf{R}_1^\mathsf{T} \overline{\mathbf{x}},$$

as can be easily derived by multiplying the first equation first *from the left* with the matrix $\mathbf{R}_1^\mathsf{T} = \mathbf{R}_1^{-1}$, then with the matrix $\mathbf{R}_2^\mathsf{T} \mathbf{M}_1$ and the matrix $\mathbf{R}_3^\mathsf{T}$, and finally by moving $\overline{\mathbf{X}}_0$ to the other side.

$\mathbf{R}_3$ turns the co-ordinate frame around the z axis from the Greenwich meridian to the local meridian of the observation site, by the angle Λ:

$$\mathbf{R}_3 = \begin{bmatrix} \cos\Lambda & +\sin\Lambda & 0 \\ -\sin\Lambda & \cos\Lambda & 0 \\ 0 & 0 & 1 \end{bmatrix}. \tag{9.2}$$

Viewed from the direction of the z axis we see (Figure 9.5) that

$$x' = x\cos\Lambda + y\sin\Lambda,$$
$$y' = -x\sin\Lambda + y\cos\Lambda.$$

The correct algebraic signs should always be verified with a sketch! See Figure 9.5. Also the direction conventions of different countries may be different.

$\mathbf{R}_2^* \overset{\text{def}}{=} \mathbf{M}_1\mathbf{R}_2$ turns the co-ordinate frame around the y axis in such a way, that the z axis will point, instead of to the celestial North Pole, to the zenith. The rotation around the y axis according to a right-handed corkscrew rule by an angle α is of the form

$$\mathbf{R}_2\left(\alpha\right) = \begin{bmatrix} \cos\alpha & 0 & -\sin\alpha \\ 0 & 1 & 0 \\ \sin\alpha & 0 & \cos\alpha \end{bmatrix}.$$

In the literature, one often finds a rotation matrix for turning *points*, not co-ordinate axes, around the given axis. Turning points corresponds to turning co-ordinate axes in the opposite direction, i.e., by a negative angle.

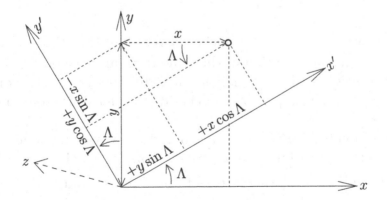

Figure 9.5: Verifying the algebraic signs of the coefficients of the rotation. The z axis is pointing out of the paper towards the reader.

Now, the rotation angle needed is $\alpha = 90° - \Phi$. Immediately after, the x axis is *mirrored*:

$$\mathbf{R}_2^* = \mathbf{M}_1 \mathbf{R}_2 = \begin{bmatrix} -1 & 0 & 0 \\ 0 & 1 & 0 \\ 0 & 0 & 1 \end{bmatrix} \cdot \begin{bmatrix} \sin\Phi & 0 & -\cos\Phi \\ 0 & 1 & 0 \\ \cos\Phi & 0 & \sin\Phi \end{bmatrix} =$$

$$= \begin{bmatrix} -\sin\Phi & 0 & \cos\Phi \\ 0 & 1 & 0 \\ \cos\Phi & 0 & \sin\Phi \end{bmatrix}. \tag{9.3}$$

$\mathbf{R}_1$ turns the co-ordinate frame around the new z axis, i.e., the vertical axis, by an amount A_0, after which the x axis points to the azimuth of the instrument's horizontal circle:

$$\mathbf{R}_1 = \begin{bmatrix} \cos A_0 & +\sin A_0 & 0 \\ -\sin A_0 & \cos A_0 & 0 \\ 0 & 0 & 1 \end{bmatrix}.$$

9.5 Solving the parameters of a three-dimensional Helmert transformation

As already shown in Section 5.1 on page 54, the three-dimensional Helmert transformation conforms to Equation 5.2 presented on page 55 in the case of small rotation angles:

$$\begin{bmatrix} x^{(2)} \\ y^{(2)} \\ z^{(2)} \end{bmatrix} = (1+\mu) \begin{bmatrix} 1 & \epsilon_z & -\epsilon_y \\ -\epsilon_z & 1 & \epsilon_x \\ \epsilon_y & -\epsilon_x & 1 \end{bmatrix} \cdot \begin{bmatrix} x^{(1)} \\ y^{(1)} \\ z^{(1)} \end{bmatrix} + \begin{bmatrix} \Delta x \\ \Delta y \\ \Delta z \end{bmatrix}. \tag{9.4}$$

Here, $\begin{bmatrix} \Delta x & \Delta y & \Delta z \end{bmatrix}^\mathsf{T}$ is the translation vector of the origin, μ is the scale distortion — i.e., the deviation of the scale from unity — and $(\epsilon_x, \epsilon_y, \epsilon_z)$ are the (small) rotation angles around the three different axes. Thus one obtains *seven* parameters. The superscripts (1) and (2) refer to the old and the new system.

This Equation 9.4 may be rearranged and linearized in the following way, making use of the approximations $\mu\epsilon_x \approx \mu\epsilon_y \approx \mu\epsilon_z \approx 0$ and substituting for the vector $\begin{bmatrix} x^{(1)} & y^{(1)} & z^{(1)} \end{bmatrix}^\mathsf{T}$ the vector of approximate values $\begin{bmatrix} x^0 & y^0 & z^0 \end{bmatrix}^\mathsf{T}$.

This is allowed, because μ and the angles ϵ are all assumed to be small.

$$\begin{bmatrix} x^{(2)} - x^{(1)} \\ y^{(2)} - y^{(1)} \\ z^{(2)} - z^{(1)} \end{bmatrix} \approx \begin{bmatrix} \mu & \epsilon_z & -\epsilon_y \\ -\epsilon_z & \mu & \epsilon_x \\ \epsilon_y & -\epsilon_x & \mu \end{bmatrix} \begin{bmatrix} x^{(1)} \\ y^{(1)} \\ z^{(1)} \end{bmatrix} + \begin{bmatrix} \Delta x \\ \Delta y \\ \Delta z \end{bmatrix} \approx$$

$$\approx \begin{bmatrix} \mu & \epsilon_z & -\epsilon_y \\ -\epsilon_z & \mu & \epsilon_x \\ \epsilon_y & -\epsilon_x & \mu \end{bmatrix} \begin{bmatrix} x^0 \\ y^0 \\ z^0 \end{bmatrix} + \begin{bmatrix} \Delta x \\ \Delta y \\ \Delta z \end{bmatrix}.$$

Rearranging terms yields

$$\begin{bmatrix} x_i^{(2)} - x_i^{(1)} \\ y_i^{(2)} - y_i^{(1)} \\ z_i^{(2)} - z_i^{(1)} \end{bmatrix} = \begin{bmatrix} x_i^0 & 0 & -z_i^0 & +y_i^0 & 1 & 0 & 0 \\ y_i^0 & +z_i^0 & 0 & -x_i^0 & 0 & 1 & 0 \\ z_i^0 & -y_i^0 & +x_i^0 & 0 & 0 & 0 & 1 \end{bmatrix} \begin{bmatrix} \mu \\ \epsilon_x \\ \epsilon_y \\ \epsilon_z \\ \Delta x \\ \Delta y \\ \Delta z \end{bmatrix}. \qquad (9.5)$$

To this Equation 9.5 we have added, for the sake of generality, the point index i, $i = 1, \ldots, n$. In this case, the number of points is n and the number of observations (co-ordinate differences available for the computation of transformation parameters) is $3n$. Thus we obtain a full *system of observation equations*:

$$\begin{bmatrix} x_1^{(2)} - x_1^{(1)} \\ y_1^{(2)} - y_1^{(1)} \\ z_1^{(2)} - z_1^{(1)} \\ \hline \vdots \\ \hline x_i^{(2)} - x_i^{(1)} \\ y_i^{(2)} - y_i^{(1)} \\ z_i^{(2)} - z_i^{(1)} \\ \hline \vdots \\ \hline x_n^{(2)} - x_n^{(1)} \\ y_n^{(2)} - y_n^{(1)} \\ z_n^{(2)} - z_n^{(1)} \end{bmatrix} = \begin{bmatrix} x_1^0 & 0 & -z_1^0 & +y_1^0 & 1 & 0 & 0 \\ y_1^0 & +z_1^0 & 0 & -x_1^0 & 0 & 1 & 0 \\ z_1^0 & -y_1^0 & +x_1^0 & 0 & 0 & 0 & 1 \\ \vdots & \vdots & \vdots & \vdots & \vdots & \vdots & \vdots \\ x_i^0 & 0 & -z_i^0 & +y_i^0 & 1 & 0 & 0 \\ y_i^0 & +z_i^0 & 0 & -x_i^0 & 0 & 1 & 0 \\ z_i^0 & -y_i^0 & +x_i^0 & 0 & 0 & 0 & 1 \\ \vdots & \vdots & \vdots & \vdots & \vdots & \vdots & \vdots \\ x_n^0 & 0 & -z_n^0 & +y_n^0 & 1 & 0 & 0 \\ y_n^0 & +z_n^0 & 0 & -x_n^0 & 0 & 1 & 0 \\ z_n^0 & -y_n^0 & +x_n^0 & 0 & 0 & 0 & 1 \end{bmatrix} \begin{bmatrix} \mu \\ \epsilon_x \\ \epsilon_y \\ \epsilon_z \\ \Delta x \\ \Delta y \\ \Delta z \end{bmatrix}. \qquad (9.6)$$

The system 9.6 of observation equations may be written symbolically:

$$\underline{\ell} = \mathbf{A}\hat{\mathbf{x}}.$$

On the left-hand side, we have an abstract vector made up of observations $\underline{\ell}$, on the right-hand side, the so-called *design matrix* $\mathbf{A}$ of the problem, multiplied with the vector $\widehat{\mathsf{x}} = \begin{bmatrix} \mu & | & \epsilon_x & \epsilon_y & \epsilon_z & | & \Delta x & \Delta y & \Delta z \end{bmatrix}^{\mathsf{T}}$, *the vector of unknowns.*

Generally, there are more observations than unknowns. This means that there are small discrepancies among the observation equations (Equations 9.6), caused by measurement errors, which should be removed in some sensible way. Let us extend the system of equations:

$$\underline{\ell} + \underline{v} = \mathbf{A}\widehat{\mathsf{x}} \tag{9.7}$$

in which $\underline{v}$ is the vector of *residuals*. The residuals cause the observation equations to agree exactly also in the case where there are more observations than unknowns and the observations contain measurement uncertainty. The best solution for the unknowns $\widehat{\mathsf{x}}$ is now the *least-squares solution*, which minimizes the sum of squared residuals. This is again one of the great discoveries of Gauss.

The system of equations contains seven unknowns $\widehat{\mathsf{x}}$ on the right-hand side. Solving the system by the method of least squares will succeed if the co-ordinates (x, y, z) are given both in the old (1) and in the new (2) system for at least three points. Then, the vector of observations $\underline{\ell}$ has nine elements or observations. In fact, already two points and one co-ordinate of a third point would be enough for this, but the redundant observations make the solution more precise.

Example. In two dimensions, Equation 9.5 simplifies as follows:

$$\begin{bmatrix} x_i^{(2)} - x_i^{(1)} \\ y_i^{(2)} - y_i^{(1)} \end{bmatrix} = \begin{bmatrix} x_i^0 & -y_i^0 & | & 1 & 0 \\ y_i^0 & x_i^0 & | & 0 & 1 \end{bmatrix} \begin{bmatrix} \mu \\ \theta \\ \hline \Delta x \\ \Delta y \end{bmatrix}. \tag{9.8}$$

This may again be summarized in the form $\underline{\ell} + \underline{v} = \mathbf{A}\widehat{\mathsf{x}}$ (residuals added), in which

$$\mathbf{A} = \begin{bmatrix} x_i^0 & -y_i^0 & | & 1 & 0 \\ y_i^0 & x_i^0 & | & 0 & 1 \end{bmatrix}.$$

The least-squares solution of this for the unknowns

$$\widehat{\mathsf{x}} = \begin{bmatrix} \mu & \theta & | & \Delta x & \Delta y \end{bmatrix}^{\mathsf{T}}$$

is now obtained by solving the *normal equations*

$$\mathbf{A}^{\mathsf{T}}\mathbf{A}\widehat{\mathsf{x}} = \mathbf{A}^{\mathsf{T}}\underline{\ell}. \tag{9.9}$$

This system of equations to be solved has as many unknowns — four — as there are equations.

The matrix $\mathbf{N} = \mathbf{A}^\mathsf{T}\mathbf{A}$ is a square matrix of size 4×4, symmetric and positive definite. It is

$$
\mathbf{N} =
\left[
\begin{array}{cc|cc}
\sum_{i=1}^{n}\left((x_i^0)^2 + (y_i^0)^2\right)\} & 0 & \sum_{i=1}^{n} x_i^0 & \sum_{i=1}^{n} y_i^0 \\
0 & \sum_{i=1}^{n}\left((x_i^0)^2 + (y_i^0)^2\right) & -\sum_{i=1}^{n} y_i^0 & \sum_{i=1}^{n} x_i^0 \\
\hline
\sum_{i=1}^{n} x_i^0 & -\sum_{i=1}^{n} y_i^0 & n & 0 \\
\sum_{i=1}^{n} y_i^0 & \sum_{i=1}^{n} x_i^0 & 0 & n
\end{array}
\right].
$$

According to this, the normal equations, Equations 9.9, given above are now

$$
\left[
\begin{array}{cc|cc}
\sum\left((x_i^0)^2 + (y_i^0)^2\right) & 0 & \sum x_i^0 & \sum y_i^0 \\
0 & \sum\left((x_i^0)^2 + (y_i^0)^2\right) & -\sum y_i^0 & \sum x_i^0 \\
\hline
\sum x_i^0 & -\sum y_i^0 & n & 0 \\
\sum y_i^0 & \sum x_i^0 & 0 & n
\end{array}
\right]
\cdot
\left[
\begin{array}{c}
\mu \\ \theta \\ \hline \Delta x \\ \Delta y
\end{array}
\right]
= \mathbf{A}^\mathsf{T}\ell.
$$

From these equations, the unknowns $\mu, \theta, \Delta x, \Delta y$ may be solved uniquely.

If we choose for the origin of the (x, y) co-ordinate frame the common *centre of mass* of the points,

$$
\left[
\begin{array}{c} x \\ y \end{array}
\right]_{\text{com}}
=
\left[
\begin{array}{c} x_{\text{com}} \\ y_{\text{com}} \end{array}
\right]
\overset{\text{def}}{=\!=}
\left[
\begin{array}{c}
\frac{1}{n}\sum_{i=1}^{n} x_i^0 \\
\frac{1}{n}\sum_{i=1}^{n} y_i^0
\end{array}
\right],
$$

the system of equations is simplified as follows:

$$
\left[
\begin{array}{cc|cc}
\sum\left((x_i^0)^2 + (y_i^0)^2\right) & 0 & 0 & 0 \\
0 & \sum\left((x_i^0)^2 + (y_i^0)^2\right) & 0 & 0 \\
\hline
0 & 0 & n & 0 \\
0 & 0 & 0 & n
\end{array}
\right]
\cdot
\left[
\begin{array}{c}
\mu \\ \theta \\ \hline \Delta x \\ \Delta y
\end{array}
\right]
= \mathbf{A}^\mathsf{T}\ell. \qquad (9.10)
$$

From this, we may infer that:

- In centre-of-mass co-ordinates we may solve the parameters each separately, when the elements of $\mathbf{A}^\mathsf{T}\ell$ have been computed.

- The precision of determination of both scale distortion μ and rotation angle θ improves, when the distance $\sqrt{(x_i^0)^2 + (y_i^0)^2}$ of the points from the origin, i.e., from their common centre of mass, increases. What this means is that, for a good determination of the transformation parameters, the points should be as "spread out" as possible, toward every edge and corner of the area of study.

- The precision of the translation parameters Δx and Δy again increases with the number of points n. The spatial distribution of these points has no impact here.

Formally, we may say that the inverse matrix $\mathbf{N}^{-1}$ describes the statistical uncertainty of the unknowns:

$$\mathbf{N}^{-1} = \left[\begin{array}{cc|cc} \left(\sum \left((x_i^0)^2 + (y_i^0)^2\right)\right)^{-1} & 0 & 0 & 0 \\ 0 & \left(\sum \left((x_i^0)^2 + (y_i^0)^2\right)\right)^{-1} & 0 & 0 \\ \hline 0 & 0 & n^{-1} & 0 \\ 0 & 0 & 0 & n^{-1} \end{array} \right]$$

$$= \frac{1}{\sigma_0^2} \left[\begin{array}{cc|cc} \sigma_\mu^2 & 0 & 0 & 0 \\ 0 & \sigma_\theta^2 & 0 & 0 \\ \hline 0 & 0 & \sigma_{\Delta x}^2 & 0 \\ 0 & 0 & 0 & \sigma_{\Delta y}^2 \end{array} \right],$$

in which σ_μ is the mean error of the parameter μ, σ_θ that of the parameter θ, and so on. Here, σ_0 represents the statistical uncertainty of a single point co-ordinate. This is often called the *mean error of unit weight*.

9.6   Variants of the Helmert transformation

The datum transformation described above, Equation 9.4 on page 118, is generally used in a geocentric co-ordinate frame. Then, the co-ordinates of place, X, Y, Z, are written as capital letters as follows:

$$\overline{\mathbf{X}'} = (1 + \mu)\mathbf{R}\overline{\mathbf{X}} + \Delta\overline{\mathbf{X}}. \tag{9.11}$$

The Helmert transformation applied geocentrically according to Equation 9.11 is called the Burša[3]–Wolf[4] presentation form. In the transformation, both the rotation $\mathbf{R}$ and the scaling $1 + \mu$ take place with respect to the Earth's centre of mass after which the translation $\Delta\overline{\mathbf{X}}$ is carried out.

In local applications, a more well-behaved variant of the Helmert transformation is the Molodenskii[5]–Badekas[6] transformation, in which rotation and

---

[3]Milan Burša (1929–) is a Czech astronomer, geodynamicist and geodesist.

[4]Helmut Wolf (1910–1994) was a German geodesist better known as the developer of the Helmert–Wolf blocking method. The method (Wolf, 1978) is useful for the adjustment of extensive geodetic networks.

[5]Mikhail Sergeevich Molodenskii (1909–1991) was a great Russian researcher into the Earth's gravity field.

[6]John Badekas was a Greek geodesist and photogrammetrist working also in the United States.

scaling take place with respect to the centre of mass of the *whole point field* $\mathbf{X}_i, i = 1, \ldots, n$:

$$\mathbf{X}_{\text{com}} = \frac{1}{n} \sum_{i=1}^{n} \mathbf{X}_i.$$

In this case, the translation vector describes the shifting of this centre of mass:

$$\overline{\mathbf{X}}' = \overline{\mathbf{X}}_{\text{com}} + (1 + \mu)\, \mathbf{R} \left(\overline{\mathbf{X}} - \overline{\mathbf{X}}_{\text{com}} \right) + \widetilde{\Delta \mathbf{X}}.$$

The matrix $\mathbf{R}$ and the scale distortion μ are identical with those of the Burša–Wolf presentation form; however, $\widetilde{\Delta \mathbf{X}} \neq \Delta \mathbf{X}$. Rewrite this:

$$\overline{\mathbf{X}}' = (1 + \mu)\mathbf{R}\overline{\mathbf{X}} + \overline{\mathbf{X}}_{\text{com}} - (1 + \mu)\, \mathbf{R}\overline{\mathbf{X}}_{\text{com}} + \widetilde{\Delta \mathbf{X}},$$

from which, by comparing with Equation 9.11, the following relationship between the two translation vectors is obtained:

$$\Delta \overline{\mathbf{X}} = \overline{\mathbf{X}}_{\text{com}} - (1 + \mu)\, \mathbf{R}\overline{\mathbf{X}}_{\text{com}} + \widetilde{\Delta \overline{\mathbf{X}}},$$

i.e.,

$$\Delta \overline{\mathbf{X}} \approx \widetilde{\Delta \overline{\mathbf{X}}} - \begin{bmatrix} \mu & \epsilon_z & -\epsilon_y \\ -\epsilon_z & \mu & \epsilon_x \\ \epsilon_y & -\epsilon_x & \mu \end{bmatrix} \mathbf{X},$$

again assuming that $\mu, \epsilon_x, \epsilon_y, \epsilon_z$ are small.

If the centre of mass $\overline{\mathbf{X}}_{\text{com}} = 0$, then $\Delta \overline{\mathbf{X}} = \widetilde{\Delta \overline{\mathbf{X}}}$, i.e., Burša–Wolf is again the same as Molodenskii–Badekas, in which the centre of mass of the point field is in the centre of mass of the Earth, i.e., the origin of the X, Y, Z system.

The advantage of the Molodenskii–Badekas presentational form is, that, at the point field, the translation and rotation are almost independent of each other. This is seen when one solves for the unknown parameters from the co-ordinates of the point field given in both datums: then the correlations between the translation and rotation parameters vanish.

On the other hand, we have here the down-side that the optimality only works in the area of the point field and it does not apply globally.

9.7 The geodetic forward or inverse problem with rotation matrices

The geodetic forward and inverse problems can be solved three-dimensionally without using the geometry of the ellipsoid of revolution. The idea is based on the three-dimensional co-ordinates of the point or points, e.g., in the form (φ, λ, h) on some reference ellipsoid; and given, or to be computed, is the azimuth, elevation angle and distance from the first point to the second point. In the forward problem, to be computed are the geodetic co-ordinates on the ellipsoid (φ, λ, h) of the second point.

9.7.1 The geodetic forward problem

Given the geodetic co-ordinates of point A, $(\varphi_A, \lambda_A, h_A)$, and, as seen from point A, the azimuth A_{AB} and distance s_{AB} of another point B, and either the elevation angle η_{AB} or the zenith angle $\zeta_{AB} \stackrel{\text{def}}{=} 90° - \eta_{AB}$.

To be computed are now the co-ordinates $(\varphi_B, \lambda_B, h_B)$ of point B. This is done as follows:

1. Convert the provided local, A-topocentric co-ordinates of point B, i.e., $(A_{AB}, s_{AB}, \zeta_{AB})$, into rectangular co-ordinates:

$$\overline{\mathbf{x}}_{AB} \stackrel{\text{def}}{=} \begin{bmatrix} x_{AB} \\ y_{AB} \\ z_{AB} \end{bmatrix} = s_{AB} \begin{bmatrix} \cos A_{AB} \sin \zeta_{AB} \\ \sin A_{AB} \sin \zeta_{AB} \\ \cos \zeta_{AB} \end{bmatrix}.$$

2. Transform, using rotation matrices, these rectangular co-ordinate differences into geocentric co-ordinate *differences*:

$$\overline{\mathbf{X}}_{AB} = \mathbf{R}_3^{\mathsf{T}} \mathbf{R}_2^{\mathsf{T}} \overline{\mathbf{x}}_{AB},$$

in which $\mathbf{R}_3$ and $\mathbf{R}_2$ are already given, see Equations 9.2 on page 117 and 9.3 on page 118.

3. Add to these differences the geocentric co-ordinates of point A, yielding the geocentric co-ordinates of point B:

$$\overline{\mathbf{X}}_B = \overline{\mathbf{X}}_A + \overline{\mathbf{X}}_{AB},$$

in which

$$\overline{\mathbf{X}}_A = \begin{bmatrix} (N(\varphi_A) + h_A) \cos \varphi_A \cos \lambda_A \\ (N(\varphi_A) + h_A) \cos \varphi_A \sin \lambda_A \\ (N(\varphi_A)(1 - e^2) + h_A) \sin \varphi_A \end{bmatrix}.$$

4. Convert the geocentric co-ordinates of B obtained back to ellipsoidal form, following Section 8.10 on page 101:

$$\overline{\mathbf{X}}_B \longrightarrow (\varphi_B, \lambda_B, h_B).$$

9.7.2 The geodetic inverse problem

Given are the ellipsoidal or geodetic co-ordinates of two points $(\varphi_A, \lambda_A, h_A)$ and $(\varphi_B, \lambda_B, h_B)$. To be computed are the topocentric spherical co-ordinates of point B, the azimuth A_{AB}, the zenith angle ζ_{AB}, and the distance s_{AB}.

1. Convert the geodetic co-ordinates of A and B into geocentric rectangular co-ordinates: $\overline{\mathbf{X}}_A$, $\overline{\mathbf{X}}_B$.

2. Calculate the relative location vector, or vector of co-ordinate differences, between the points,
$$\overline{\mathbf{X}}_{AB} = \overline{\mathbf{X}}_B - \overline{\mathbf{X}}_A.$$

3. Transform this vector into the topocentric rectangular system of point A:
$$\overline{\mathbf{x}}_{AB} = \mathbf{R}_2 \mathbf{R}_3 \overline{\mathbf{X}}_{AB};$$

4. Convert to spherical co-ordinates by solving from the equation
$$\overline{\mathbf{x}}_{AB} = s_{AB} \begin{bmatrix} \cos A_{AB} \sin \zeta_{AB} \\ \sin A_{AB} \sin \zeta_{AB} \\ \cos \zeta_{AB} \end{bmatrix}$$

the co-ordinates s_{AB}, A_{AB} and ζ_{AB}. For this, the familiar arc tangent formula and the Pythagoras theorem may be used. One may avoid the quadrant problem of the arc tangent by using the half-angle formula:

$$s_{AB} = \sqrt{x_{AB}^2 + y_{AB}^2 + z_{AB}^2},$$

$$\tan A_{AB} = \frac{y_{AB}}{x_{AB}} = 2 \arctan \frac{y_{AB}}{x_{AB} + \sqrt{x_{AB}^2 + y_{AB}^2}},$$

$$\tan \zeta_{AB} = \frac{\sqrt{x_{AB}^2 + y_{AB}^2}}{z_{AB}}.$$

9.7.3 Comparison with the ellipsoidal surface solution

In Chapter 8 we discussed the geometry of the reference ellipsoid and the solution of the forward and inverse geodetic problems on the surface of the ellipsoid. The solution obtained in spherical geometry is, where azimuths are concerned, very close to the solution obtained by using the geodesic between the projection points of A and B on the surface of the ellipsoid. *These solutions are, however, not identical.* The azimuths A_{AB} and A_{BA} are so-called *normal-section azimuths*, which differ by a fraction of a second of arc from the azimuths of the geodesic, even over a distance of a hundred kilometres (Torge, 2001, page 240). The difference is minute, but not zero!

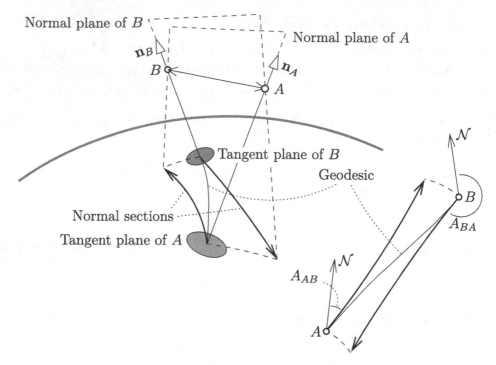

Figure 9.6: Normal sections and the geodesic on the ellipsoid. The transversal difference in direction between the ellipsoidal normals $\mathbf{n}_A$ and $\mathbf{n}_B$ has been strongly exaggerated.

In order to visualise this difference, Figure 9.6 presents the three-dimensional observation situation in points A and B with their normal sections. The ellipsoidal normals in points A and B do not lie in the same plane, and the intersection curves with the ellipsoidal surfaces differ both from each other and from the geodesic. This would be true even if points A and B were located *on* the surface of the reference ellipsoid. In the figure they are drawn, for clarity, above the ellipsoidal surface.

On the right-hand side of Figure 9.6 are shown all three curves, projected on the surface of the ellipsoid.

The distance s_{AB} is of course the length of the straight line in space, not of the geodesic. This difference is significant.

Exercises

Exercise 9 − 1:

Derive the equation corresponding to Equation 9.10 on page 121 for three dimensions.

Chapter 10

Co-ordinate reference systems

10.1 History

The grade measurement of the French Academy of Sciences, described in subsection 1.6.2 on page 13, in Lapland and Peru, was one of the first attempts to determine the precise figure of the Earth. It did not remain the last one. For example, in Eastern Europe, in the years 1816–1855, the so-called Struve[1] geodetic chain, or meridian arc, was realized, running from the Norwegian coast to the coast of the Black Sea, a total length of over 2820 km. Also the measurement of this triangle chain was intended to study the precise figure of the Earth within the European territory. For this purpose, in addition to triangulation, astronomical measurements were carried out to determine the direction of the local plumb line. The meridian arc, consisting of 258 triangles, and of which also on the present Finnish territory points have been preserved (Veriö, 1994), was declared in 2005 a UNESCO World Heritage Site, see NLS (maintained).

In a rapidly industrializing Europe, precise mapping was of vital importance for the cadastre, zoning, and land use, and for that, knowledge of the precise figure of the Earth was a just as essential problem. In many countries, national mapping projects were started, and in 1862, on the initiative of lieutenant-general Baeyer[2], a commission was formed by the name of *Mitteleuropäische Gradmessung* ("Central European Grade Measurement"), which developed over time (1919) into an organization called *International Association of Geodesy*. It is the international organization uniting the global geodetic research community.

In the early days, every country created its own geodetic base network and chose its own starting points; this is how a large number of national location and height datums were created. When, in the 20th century, international co-operation intensified, the need arose to also create uniform, initially continent-wide, geodetic datums. In Europe, the first modern such datum was probably the European Datum 1950 (ED50), which connected, then, only the triangle

---

[1]Friedrich Georg Wilhelm von Struve (1793–1864) was a German astronomer and geodesist working at the Tartu Observatory.

[2]Johann Jacob Baeyer (1794–1885) was a Prussian officer, geodesist and diplomat of science; he was the instigator and organizer of the Central European Grade Measurement.

networks of the Western European countries, in a joint computation creating a unified geodetic datum, the starting point of which was Munich in Germany. ED50 was also the starting point for the Finnish national datum, the National Map Grid Co-ordinate System (KKJ) created in 1970 — although its realization was a bit special, see Section 6.1.1 on page 67.

Precisely connecting the geodetic systems of different continents remained long impossible. Still on May 20, 1947, using a total solar eclipse, the Finnish researchers T. J. Kukkamäki[3] and R. A. Hirvonen carried out a measurement of the distance between Africa (Gold Coast) and South America (Brazil), the precision of which was a record 141 m.

Only satellites made a more accurate measurement of intercontinental distances possible. The first attempts were already undertaken in the 1960s, when geodetic satellites, such as ANNA-1B and Pageos (Passive Geodetic Satellite), were being photographed by massive cameras, applying the stellar triangulation technique of Yrjö Väisälä[4] (Kakkuri and Kivioja, 2012; Väisälä, 1946).

With these, and the still primitive satellite navigation system "Transit," the first global geodetic systems were created: the WGS (World Geodetic System) editions of 1966, 1972, and eventually 1984. Later came GPS, the Global Positioning System.

10.2   Parameters of the GRS80 system

Geodetic Reference System 1980 (GRS80) was approved by the IUGG, the International Union of Geodesy and Geophysics, in its General Assembly of 1979 in Canberra, Australia. Its defining parameters (e.g., Heikkinen, 1981) are presented in Table 10.1.

Of the parameters, surprisingly only one is geometric (the semi-major axis a); the others are dynamical, like the dynamic flattening J_2 quantifying the flattening of the Earth's gravitational field, and the angular rate of rotation ω. Not even the semi-minor axis b is a defining parameter, and neither is the flattening f. But the other geometric and dynamic parameters can be derived directly, though the derivations are complicated.

---

[3]Tauno Johannes ("TJ") Kukkamäki (1909–1997) was a Finnish geodesist and specialist on the metrology of length, and a world famous expert on levelling refraction.

[4]Yrjö Väisälä (1891–1971), "the Wizard of Tuorla"(Niemi, 1991), was a Finnish academician in geodesy and astronomy, physicist, telescope builder, discoverer of comets and asteroids, metrologist of length, pleasure yachter and Esperantist.

Table 10.1: Defining parameters of the GRS80 system.

Quantity	Value	Explanation
a	$6378\,137\,\text{m}$	Semi-major axis
GM	$3986\,005 \cdot 10^8\,\text{m}^3/\text{s}^2$	Earth's mass$\times G$
J_2	$108\,263 \cdot 10^{-8}$	Dynamic flattening
ω	$7292\,115 \cdot 10^{-11}\,\text{rad}/\text{s}$	Angular rate of rotation

We combine, following Moritz (1992) and Heiskanen and Moritz (1967), equations 2-90 and 2-92 given in the latter source on page 73, and obtain

$$J_2 = \frac{e^2}{3}\left(1 - \frac{2}{15}\frac{me'}{q_0}\right) \implies e^2 = 3J_2 + \frac{2me'e^2}{15q_0}.$$

The definitions of the quantities e', m and q_0 appearing here may be found in Section 8.12 on page 103. The definition of parameter m is Equation 8.13 on page 105:

$$m = \frac{\omega^2 a^2 b}{GM}.$$

One obtains, using $be' = ae$:

$$e^2 = 3J_2 + \frac{4}{15}\frac{\omega^2 a^3}{GM}\frac{e^3}{2q_0}. \tag{10.1}$$

For the parameter q_0 there is Equation 8.12 on page 105, and e' may be written as a function of e:

$$e' = \frac{e}{\sqrt{1 - e^2}}.$$

Now we may compute e^2 *iteratively* using Equation 10.1, recomputing at every step $q_0\left(e'\left(e\right)\right)$. As a result, one obtains the values for the derived parameters given in Table 10.2.

Here we have used the relationships derived in Section 8.7 on page 99, Equations 8.5 and 8.6, between e^2, e'^2 and f.

Table 10.2: Derived parameters of the GRS80 system.

Quantity	Value	Explanation
$e^2 = \dfrac{a^2 - b^2}{a^2}$	$0.006\,694\,380\,022\,90$	First eccentricity squared
$e'^2 = \dfrac{a^2 - b^2}{b^2}$	$0.006\,739\,496\,775\,48$	Second eccentricity squared
b	$6356\,752.314\,140\,\text{m}$	Semi-minor axis
$\dfrac{1}{f}$	$298.257\,222\,101$	Inverse flattening

Table 10.3: Parameters of the WGS84 reference ellipsoid.

Quantity	Value	Remark
a	6378 137 m	Same as GRS80
$\frac{1}{f}$	298.257 223 563	Different!
b	6356 752.314 245 m	Difference 0.1 mm

Often one uses (a, f) together to describe the GRS80 reference ellipsoid.

The reference ellipsoid used by the GPS is the ellipsoid of the WGS84 (World Geodetic System 1984). It is in principle the same as GRS80, but, due to reasons related to numerical computational imprecision (NIMA, 3 January 2000, § 3.2.2), it has a slightly different flattening, see Table 10.3. As a result, the semi-minor axis of WGS84, its polar radius, is 0.1 mm shorter than the one from GRS80[5].

The *gravimetric parameters* of the GRS80 ellipsoid are computed using the equations of Section 8.12 on page 103, see Table 10.4.

As a check, we may still calculate Clairaut's theorem in its precise form (Heiskanen and Moritz, 1967, equation 2-75):

$$f + f^* = \frac{\omega^2 b}{\gamma_e}\left(1 + e'\frac{q_0'}{2q_0}\right).$$

10.3 GRS80 and modern co-ordinate reference systems

In Finland, like in many other countries, traditional reference systems had long been in use, back then determined based on terrestrial geodetic

---

[5]Calculation:

$$\Delta b = -\frac{\Delta f}{f}(a - b) = \frac{\Delta\left(\frac{1}{f}\right)}{\left(\frac{1}{f}\right)}(a - b) = \frac{0.000\,001\,462}{298.257\,22} \cdot (21\,384\,\text{m}) = 0.000\,105\,\text{m}.$$

Table 10.4: Gravimetric parameters for the GRS80 reference ellipsoid.

Quantity	Value	Explanation
γ_e	9.780 326 7715 m/s²	Normal gravity on the equator
γ_p	9.832 186 3685 m/s²	Normal gravity on the poles
f^*	0.005 302 440 112 29	"Gravity flattening"

measurements (Häkli et al., 2009). Without satellite positioning methods, it is not possible to achieve a truly geocentric co-ordinate reference system.

The old European system ED50 was created in the 1950s by uniting the geodetic triangle networks of Western Europe. As the computation surface, the Hayford[6] or International Ellipsoid was chosen, which the IUGG had accepted already in 1924, and which differs significantly from the GRS80 ellipsoid. Also the triangulation network of Finland was computed on the Hayford ellipsoid, and the KKJ created in 1970 was based on this computation.

The system is not even approximately geocentric: modern GPS measurements have shown that the nominal centre of the ellipsoid of computation used was located over a hundred metres from the Earth's centre of mass! Furthermore, the system contained substantial distortions, e.g., between Northern and Southern Finland, and also the scale was not very good.

The new satellite-based systems, to which a general transition has taken place, are much more homogeneous in their precision, internationally consistent, and *genuinely geocentric*: the deviations from the Earth's centre of mass are of order a few centimetres.

The main differences between the traditional and modern co-ordinate reference systems used in Finland are thus:

1. the reference ellipsoid used: the International Ellipsoid (Hayford) of 1924 vs. the GRS80 ellipsoid

2. a realization based on terrestrial measurements vs. on satellite measurements (and space geodetic measurements more generally)

3. non-geocentric (level of discrepancies of order $100\,\mathrm{m}$) vs. geocentric (discrepancies of order several cm).

In order to determine the figure of the reference ellipsoid, *two quantities* are needed, for example, the semi-major axis or equatorial radius a and the flattening f. The values for the Hayford and GRS80 ellipsoids are

○ International Ellipsoid 1924: $a = 6378\,388\,\mathrm{m}$, $f = \frac{1}{297}$

○ GRS80: $a = 6378\,137\,\mathrm{m}$, $f = \frac{1}{298.257\,222\,101}$.

The geodetic research community uses in its published products exclusively parameters of the GRS80 reference ellipsoid. The parameters of the system are found in Tables 10.1 and 10.2. About these products, like the ITRS reference system (International Terrestrial Reference System) and its realizations, we shall tell more later on. In precise geodetic work, these are always used, though often one also speaks semi-carelessly of WGS84 co-ordinates.

---

[6]John Fillmore Hayford (1868 – 1925) was an American geophysicist and geodesist.

10.4   Co-ordinate reference systems and their realizations

10.4.1   Terminology

Both nationally and internationally, there is a somewhat varying terminology in use for referring to the *realizations* of co-ordinate reference systems. This means that a description in principle exists of how a co-ordinate reference system is created: its fundamental parameters and so on. This description is called a *reference system*. The actual system is then realized by practical measurements. In this way, a point set is created having known co-ordinate values. This in turn is called a *reference frame*, and the process itself a *realization* of the system. In geodesy, an often used term is also *datum*, i.e., the way in which a set of points has been connected to the terrain by measurements, often by fixing some starting points to conventionally agreed co-ordinate values.

The following terminology is used:

o International Organization for Standardization (ISO):

   co-ordinate reference system / co-ordinate system

o International Earth Rotation and reference Systems Service (IERS):

   reference system / reference frame

o in Finnish:

   koordinaattijärjestelmä / koordinaatisto[7] (JUHTA, 2016a, JHS196).

Of these terms, *the latter* is used for a system that has been realized in the terrain by means of real measurements, resulting in co-ordinate values for the measurement points or stations. Thus, we have a *realization*. In this process, also one or more *datum points* are defined. The co-ordinates of these points are fixed, as part of the datum definition, to agreed — i.e., to some extent arbitrary, not measured — values.

Of the terms, the *former* refers on an abstract level to the definition of the co-ordinate reference system, like the choice of reference ellipsoid and origin — e.g., the centre of mass of the Earth — as well as the orientation of the axes.

10.4.2   The practical viewpoint

The computation of the various realizations of the ITRS is carried out by using a broad range of space geodetic measurements: one uses GPS, the

---

[7]The Finnish geodetic terminology has recently been revised, as is described in the new JHS publications 196 and 197. The new terminology agrees better with international practices.

Russian GLONASS system, very long baseline interferometry (VLBI) with radio telescopes, as well as laser ranging measurements to satellites and to the Moon.

Both WGS84 and ITRS are *realized* by computing co-ordinates and velocity vectors for the vertices of a polyhedron, i.e., the various stations on the Earth's surface.

The fundamental properties of the systems are (IERS, 1996, 2003, IERS conventions):

o They are *geocentric*, i.e., the origin of the co-ordinates, and thus also of the reference ellipsoid used, is the centre of mass of the Earth (included in the mass of the Earth are also the oceans and the atmosphere, but *not* the Moon!). The geocentricity is realized by using satellite observations; the equations of orbital motion of the satellites are given in a co-ordinate frame that is implicitly geocentric.

o The *scale* comes directly from the SI system, and is realized by using electromagnetic range measurements. The speed of propagation of electromagnetic waves in vacuum is conventionally $299\,792\,458$ ^m/s. In this way, distances are measured by measuring time differences using an atomic clock, an extremely precise method.

o The *orientation*: for the direction of the Earth's rotation axis the conventional international origin (CIO) was originally used, i.e., the average direction of the axis over the years $1900-1905$, and the direction of the Greenwich meridional plane, i.e., essentially the direction of the plumb line at the location of the Greenwich transit circle. Nowadays, the orientation is maintained by the IERS using VLBI and GPS, and, although the formal definition has changed, continuity is preserved.

o The current *definition* is based on the definition of the pole and the direction of the zero meridian according to the BIH (*Bureau International de l'Heure*[8]) agreement of 1984. Together, the co-ordinates X, Y and Z form a right-handed system.

Because the measurement technologies used are extremely precise, also the various motions and deformations of the Earth — like tidal phenomena — must be taken into account by models that are both precise and correct. On this level of precision, already plate tectonics invalidates the assumption that station co-ordinates are constant. The assumption made instead is, that they are linear functions of time (IERS, 2003) equation (12):

$$\mathbf{X}(t) = \mathbf{X}_0 + \dot{\mathbf{X}} \cdot (t - t_0),$$

---

[8]International Time Bureau, the predecessor of IERS.

where t_0 is the reference epoch. The parameter solution for every station obtained in this computation is thus the pair of estimates $\left(\widehat{\mathbf{X}}_0, \widehat{\dot{\mathbf{X}}}\right)$.

The existing realizations of the ITRS are: ITRF88, 89, 90, 91, 92, 93, 94, 96, 97, ITRF2000, ITRF2005, ITRF2008 and ITRF2014. One may ask why, always after a few years — and in the beginning even at yearly intervals — a new realization of the system is computed. There are several reasons:

o Observational data, from the global network of continuously operating stations, is being collected more and more of all the time, i.e., recomputation will produce better accuracies.

o The technologies of both the satellite systems and the receivers are developing, and also the number and geographical distribution of stations are improving.

o The models used in the computations are improving, thanks to scientific research.

o The realizations have a "best before" date. For example, ITRF96 has been computed from data that extends up to the year 1997. GPS data started to become available for inclusion in 1991, VLBI data was available and included in the computations already from 1979, and satellite laser data from 1976 (Boucher et al., 1998). However, high quality, abundant data was available only from a period of five years. Extrapolation from this data far into the future, e.g., to the year 2012, would not be a good idea.

Finally, the precise orbital elements distributed over the Internet by the IGS, the International GNSS Service, are computed using the data produced by the GNSS tracking stations in Figure 10.1. This network is largely the same as that on which also the computation of the ITRS realizations is based. Also in Finland there is a tracking station used by the IGS, Metsähovi. The orbital elements are always referred to the last valid ITRS realization, i.e., the latest available ITRF.

10.4.3   The ITRS/ETRS systems and their realizations

All mentioned systems and realizations fall under the responsibility of the IERS, a service of the international geodetic scientific community organized in the International Association of Geodesy (IAG). There is quite a bit of acronym soup:

o "I" International

o "E" European

o "T" Terrestrial

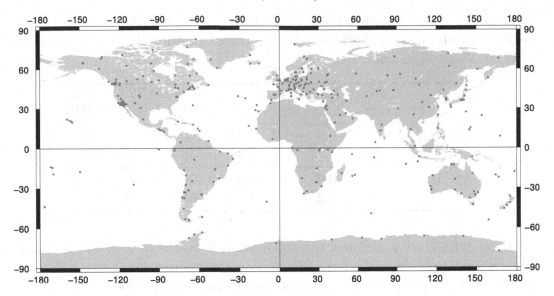

Figure 10.1: The IGS tracking stations, status 2010 (data © IGS).

○ "S" System, and

○ "F" Frame.

The letter S, *System*, refers to the principles guiding the definition of the reference system before the realization is carried out. There is only a single ITRS, but there are multiple ITRFs (*Frames*), which are its *realizations*, i.e., co-ordinate solutions computed from real measurements for networks on the Earth's surface.

The same applies also to the pair of acronyms ETRS/ETRF standing for the corresponding concepts on the European territory, in which the motion of the Eurasian tectonic plate (about 2.5 cm per year) has been computationally eliminated. In this way, co-ordinates are obtained that are constant, the use of which in geographic information systems is as straightforward as possible.

In Europe today, the system ETRS89 is in general use, even as a reference system standard accepted at the EU level. The system was initiated by the *Reference Frame Subcommission for Europe* (EUREF subcommission) of the IAG. This is also the origin of the name "EUREF system."

In all European countries, national realizations of this system have been created by measurement campaigns. The realization used in Finland is called EUREF-FIN. ETRS89 has been defined in such a way, that for the epoch 1989.0 — i.e., the start of the year 1989 — it is identical to the ITRS. The time dependence of co-ordinates caused by the plate tectonics of the Eurasian plate has been computationally removed. Therefore, the co-ordinates of the ITRF and the ETRF realizations differ from each other for epochs later than 1989.0.

In other parts of the world, similar approaches have been adopted. For example, in North America, the datum NAD 83 (North American Datum 1983)

Table 10.5: Transformation parameters from the ITRF2008 system to the ITRF2005 system, epoch 2005.0, see ITRF, Transformation Parameters between ITRF2005 and ITRF2008.

	$T1$ mm	$T2$ mm	$T3$ mm	D ppb	$R1$ $0.001''$	$R2$ $0.001''$	$R3$ $0.001''$
Value	−0.5	−0.9	−4.7	0.94	0.000	0.000	0.000
±	0.2	0.2	0.2	0.03	0.008	0.008	0.008
Rate	0.3	0.0	0.0	0.00	0.000	0.000	0.000
±	0.2	0.2	0.2	0.03	0.008	0.008	0.008

has been defined by a Helmert transformation from the newest ITRF frame, with linearly time dependent transformation parameters aiming to eliminate the effect of the plate motion on point co-ordinates (Snay, 2003).

10.5 Transformations between different ITRFs

Because every ITRS realization ITRF*yy* has been computed as a set of station location and velocity vectors, one must takie into account also in the transformations between them the time dependence of the transformation parameters. Thus one uses a *fourteen-parameter Helmert transformation*, in which, in addition to the seven parameter values of the standard Helmert transformation (translation vector, rotation matrix and scale correction), also their first time derivatives or rates of change are estimated as parameters.

The transformation parameters between the different ITRFs are found on the web site ITRF, Transformation Parameters.

As an example, we give here the transformation parameters from the ITRF2008 system to the ITRF2005 system, for epoch 2005.0, as they are found on that web page, see Table 10.5.

These parameters[9] should be applied in the following way:

$$
\begin{bmatrix} X \\ Y \\ Z \end{bmatrix}_{\text{ITRF2005}}(t) = \left\{ 1 + \begin{bmatrix} D & -R_3 & R_2 \\ R_3 & D & -R_1 \\ -R_2 & R_1 & D \end{bmatrix} \right\} \begin{bmatrix} X \\ Y \\ Z \end{bmatrix}_{\text{ITRF2008}}(t) +
$$
$$
+ (t - t_0) \frac{d}{dt} \begin{bmatrix} D & -R_3 & R_2 \\ R_3 & D & -R_1 \\ -R_2 & R_1 & D \end{bmatrix} \begin{bmatrix} X \\ Y \\ Z \end{bmatrix}_{\text{ITRF2008}}(t) +
$$

---

[9]Note that the IERS publications use different names for the parameters. The translation parameters $(\Delta x, \Delta y, \Delta z)$ are now $(T1, T2, T3)$; μ is D; and $(\epsilon_x, \epsilon_y, \epsilon_z)$ is now $(R1, R2, R3)$.

$$+ \begin{bmatrix} T_1 \\ T_2 \\ T_3 \end{bmatrix} + (t - t_0) \frac{d}{dt} \begin{bmatrix} T_1 \\ T_2 \\ T_3 \end{bmatrix} =$$

$$= \left\{ 1 + \begin{bmatrix} 0.94 & 0 & 0 \\ 0 & 0.94 & 0 \\ 0 & 0 & 0.94 \end{bmatrix} \cdot 10^{-9} \right\} \cdot \begin{bmatrix} X \\ Y \\ Z \end{bmatrix}_{\text{ITRF2008}} (t) +$$

$$+ \begin{bmatrix} -0.5\,\text{mm} + 0.3\,\text{mm/a}\,(t - 2005.0) \\ -0.9\,\text{mm} \\ -4.7\,\text{mm} \end{bmatrix}, \qquad (10.2)$$

using the given numerical values and forgetting about uncertainties. Here, $\frac{d}{dt}$ refers to rates of change, although only the rate of change of T_1 differs in this example from zero.

10.6 The Finnish ETRS89 realization, EUREF-FIN

The Finnish national ETRS89 realization or *datum* is called EUREF-FIN. It has been realized by means of GPS measurements, in which the continuously operating stations Metsähovi, Vaasa, Joensuu and Sodankylä belonging to the FinnRef[10] acted as datum points, i.e., they together defined the datum (Ollikainen et al., 2000).

Realizing a reference co-ordinate system on a moving crustal plate in an area where, as a consequence of post-glacial land uplift, the land continues to slowly rise, is not quite simple. Of the work of the IAG's EUREF subcommission a large part consists of providing guidance on how European countries should proceed in order to establish EUREF co-ordinates for the whole continent, co-ordinates that are as homogeneous, intercompatible and practically useful as possible.

The Finnish co-ordinates computed from the GPS measurements, obtained in the ITRF96 co-ordinate frame for measurement epoch 1997.0, should first be *transformed* using the procedure described in the memo Boucher and Altamimi (1995), in order to obtain co-ordinates in a system called ETRF96. Although the co-ordinate *system* is ETRS89, the *epoch of its realization* is nevertheless 1996, and the central epoch of the observations is again 1997.0! The latter circumstance is significant especially in the Fennoscandian area, because the definition of ETRS89 does not consider the Fennoscandian land uplift.

---

[10]FinnRef is a trademark of the Finnish National Land Survey.

The transformation formula used is of the form (Häkli et al., 2009):

$$\mathbf{X}^{\mathrm{E}}(t_{\mathrm{C}}) = \mathbf{X}_{yy}^{\mathrm{I}}(t_{\mathrm{C}}) + \mathbf{T}_{yy} + \begin{bmatrix} 0 & -\dot{R}_3 & \dot{R}_2 \\ \dot{R}_3 & 0 & -\dot{R}_1 \\ -\dot{R}_2 & \dot{R}_1 & 0 \end{bmatrix}_{yy} \mathbf{X}_{yy}^{\mathrm{I}}(t_{\mathrm{C}}) \cdot (t_{\mathrm{C}} - 1989.0),$$

in which t_{C} is the central epoch of the observations used, and $yy = (19)96$ says that

1. The GPS material has been processed using ITRF96 (more precisely, in addition to the co-ordinates of the ground stations, also the satellite ephemeris used in the computation were in this system), and

2. The results are desired in the corresponding ETRS89 realization ETRF96.

The values $\mathbf{T}_{96}$ (translation vector) and $\dot{R}_{i,96}$ (the elements of the rotation-rate matrix) may be lifted from the memo (Boucher and Altamimi, 2008, appendix 3, tables 3 and 4).

10.7   A triangle-wise affine transformation for the Finnish territory

In Finland, the National Land Survey has developed a method to transform KKJ co-ordinates into the new co-ordinates of the ETRS-TM35FIN system and map projection. The method, which is documented in Julkisen Hallinnon Suositus (JHS, "Recommendation for Public Administration") no. 197 (JUHTA, 2016b), works in a way that the area of Finland is divided into triangles. Inside every triangle, an affine transformation is defined, the co-ordinates of which are computed from the co-ordinate differences between the two systems at the corners of the triangle using bilinear interpolation, in other words, by fitting a *plane* through these three points.

Because adjacent triangles have two common corner points or vertices, we achieve, by this method of computation, that the interpolation is *continuous* across triangle borders.

Inside every triangle, the affine transformation may be written in the form

$$x^{(2)} = \Delta x + a_1 x^{(1)} + a_2 y^{(1)}$$
$$y^{(2)} = \Delta y + b_1 x^{(1)} + b_2 y^{(1)}$$

in which $\left(x^{(2)}, y^{(2)}\right)$ are the co-ordinates of the point in the new ETRS-TM35FIN, and $\left(x^{(1)}, y^{(1)}\right)$ are the co-ordinates of the same point in the old

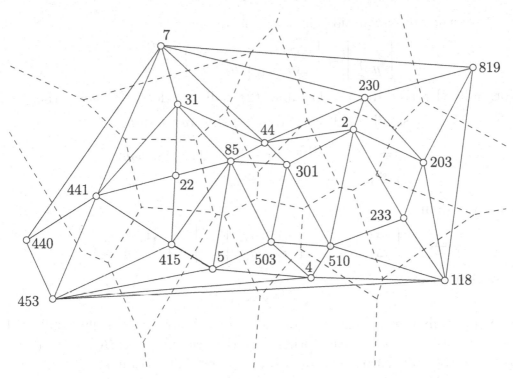

Figure 10.2: The local densification of the national triangulation for the city of Lappeenranta (Vermeer et al., 2004). Solid lines represent the Delaunay triangulation, dashed lines represent the corresponding Voronoi diagram.

YKJ[11]. This transformation formula has six parameters: Δx, Δy, a_1, a_2, b_1 and b_2. When three vertices of the triangle have been given in both systems, $\left(x^{(1)}, y^{(1)}\right)$ as well as $\left(x^{(2)}, y^{(2)}\right)$, one may solve these uniquely.

The transformation function obtained is linear inside the triangles and continuous across triangle borders, but not differentiable: the scale is discontinuous at the triangle edges. Because the mapping is not conformal, the scale will also depend on the direction considered.

We may consider it a useful property of the triangulation, that it may be locally densified or improved: if better data exists for a local area — a denser point set, for which the co-ordinate pairs $\left(x^{(i)}, y^{(i)}\right)$, $i = 1, 2$ are known. Then the triangles in the area under consideration may be replaced by a larger number of smaller triangles, inside which the transformation becomes more precise. Figure 10.2 gives an example of this process in the case of the city of Lappeenranta.

---

[11]YKJ, the Uniform Co-ordinate System, i.e., the Map Grid Co-ordinate System (KKJ) for central meridian $27°$ East, as once used for the whole of Finland.

Write the above equations in vector form:

$$\begin{bmatrix} x^{(2)} \\ y^{(2)} \end{bmatrix} = \begin{bmatrix} \Delta x \\ \Delta y \end{bmatrix} + \begin{bmatrix} a_1 & a_2 \\ b_1 & b_2 \end{bmatrix} \begin{bmatrix} x^{(1)} \\ y^{(1)} \end{bmatrix}.$$

Normally the co-ordinates (1) and (2) will be close to each other, i.e., $\begin{bmatrix} \Delta x & \Delta y \end{bmatrix}^\mathsf{T}$ is small. Then we may write the *translations*

$$\delta x \stackrel{\text{def}}{=} x^{(2)} - x^{(1)} = \Delta x + (a_1 - 1)\, x^{(1)} + a_2 y^{(1)},$$
$$\delta y \stackrel{\text{def}}{=} y^{(2)} - y^{(1)} = \Delta y + b_1 x^{(1)} + (b_2 - 1)\, y^{(1)}.$$

If we now define

$$\Delta \mathbf{x} \stackrel{\text{def}}{=} \begin{bmatrix} \Delta x \\ \Delta y \end{bmatrix}, \quad \mathbf{A} = \begin{bmatrix} a_{11} & a_{12} \\ a_{21} & a_{22} \end{bmatrix} \stackrel{\text{def}}{=} \begin{bmatrix} a_1 - 1 & a_2 \\ b_1 & b_2 - 1 \end{bmatrix},$$

we obtain, in short,

$$\delta \mathbf{x} = \Delta \mathbf{x} + \mathbf{A}\mathbf{x}^{(1)}.$$

Here, generally also the elements of the matrix $\mathbf{A}$ are numerically small, if the co-ordinates are close to each other. Let the triangle be ABC. In that case, given are the translation vectors of the vertices of the triangle

$$\delta \mathbf{x}_A = \Delta \mathbf{x} + \mathbf{A}\mathbf{x}_A^{(1)},$$
$$\delta \mathbf{x}_B = \Delta \mathbf{x} + \mathbf{A}\mathbf{x}_B^{(1)},$$
$$\delta \mathbf{x}_C = \Delta \mathbf{x} + \mathbf{A}\mathbf{x}_C^{(1)}.$$

Write the equations out so that the unknowns $\Delta \mathbf{x}, \mathbf{A}$ are on the right-hand side:

$$\delta x_A = \Delta x + a_{11} x_A^{(1)} + a_{12} y_A^{(1)},$$
$$\delta y_A = \Delta y + a_{21} x_A^{(1)} + a_{22} y_A^{(1)},$$
$$\delta x_B = \Delta x + a_{11} x_B^{(1)} + a_{12} y_B^{(1)},$$
$$\delta y_B = \Delta y + a_{12} x_B^{(1)} + a_{22} y_B^{(1)},$$
$$\delta x_C = \Delta x + a_{11} x_C^{(1)} + a_{12} y_C^{(1)},$$
$$\delta y_C = \Delta y + a_{21} x_C^{(1)} + a_{22} y_C^{(1)},$$

or, in matric form,

$$\begin{bmatrix} \delta x_A \\ \delta y_A \\ \delta x_B \\ \delta y_B \\ \delta x_C \\ \delta y_C \end{bmatrix} = \begin{bmatrix} 1 & 0 & x_A^{(1)} & 0 & y_A^{(1)} & 0 \\ 0 & 1 & 0 & x_A^{(1)} & 0 & y_A^{(1)} \\ 1 & 0 & x_B^{(1)} & 0 & y_B^{(1)} & 0 \\ 0 & 1 & 0 & x_B^{(1)} & 0 & y_B^{(1)} \\ 1 & 0 & x_C^{(1)} & 0 & y_C^{(1)} & 0 \\ 0 & 1 & 0 & x_C^{(1)} & 0 & y_C^{(1)} \end{bmatrix} \begin{bmatrix} \Delta x \\ \Delta y \\ a_{11} \\ a_{21} \\ a_{12} \\ a_{22} \end{bmatrix},$$

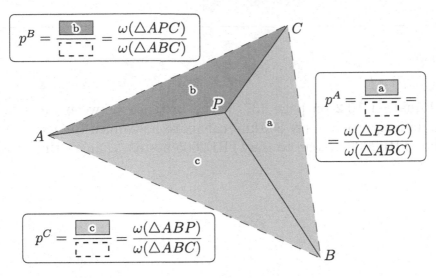

Figure 10.3: Computing barycentric co-ordinates as ratios of surface areas of two triangles.

from which all six unknowns may be solved.

Write the co-ordinates (x, y) of an arbitrary point P as follows:

$$x = p^A x_A + p^B x_B + p^C x_C,$$
$$y = p^A y_A + p^B y_B + p^C y_C,$$

with the additional condition $p^A + p^B + p^C = 1$. Then also

$$\delta x = p^A \delta x_A + p^B \delta x_B + p^C \delta x_C,$$
$$\delta y = p^A \delta y_A + p^B \delta y_B + p^C \delta y_C.$$

The triple $\left(p^A, p^B, p^C\right)$ is called the *barycentric co-ordinates* of point P. See Figure 10.3. They are found in the following way (geometrically $p^A = \frac{\omega(\triangle BCP)}{\omega(\triangle ABC)}$ etc., in which ω is the surface area of a triangle) using determinants:

$$p^A = \frac{\begin{vmatrix} x_B & x_C & x \\ y_B & y_C & y \\ 1 & 1 & 1 \end{vmatrix}}{\begin{vmatrix} x_A & x_B & x_C \\ y_A & y_B & y_C \\ 1 & 1 & 1 \end{vmatrix}}, \quad p^B = \frac{\begin{vmatrix} x_C & x_A & x \\ y_C & y_A & y \\ 1 & 1 & 1 \end{vmatrix}}{\begin{vmatrix} x_A & x_B & x_C \\ y_A & y_B & y_C \\ 1 & 1 & 1 \end{vmatrix}}, \quad p^C = \frac{\begin{vmatrix} x_A & x_B & x \\ y_A & y_B & y \\ 1 & 1 & 1 \end{vmatrix}}{\begin{vmatrix} x_A & x_B & x_C \\ y_A & y_B & y_C \\ 1 & 1 & 1 \end{vmatrix}}.$$

These equations are directly suitable for coding.

Exercises

Exercise 10 − 1:

Using Equation 10.2 on page 137, if the distance between Metsähovi and Sodankylä is approximately 800 km, calculate how much the transformation from the ITRF2008 system to the ITRF2005 system changes this distance.

Chapter 11

Co-ordinates of heaven and Earth

The Earth moves in space in several different ways. The most important of those are its orbital motion around the Sun and its rotation around its own axis. Also, the orbit and the direction of the axis of rotation are not constant, but change slowly in the course of time. The natural satellite of the Earth, the Moon, is again so massive that it affects also the motion of the Earth, causing a small monthly orbital motion of the Earth around the common centre of mass of the Earth-Moon system, or barycentre.

In Figure 11.1 we see the apparent path of the Sun in the sky, or *ecliptic*, the tilt of the Earth's axis of rotation with the plane of the ecliptic — i.e., the plane of the Earth's orbit — and how this tilt, or *obliquity*, causes that most

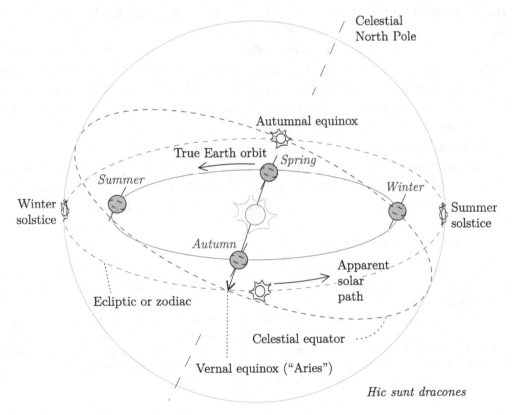

Figure 11.1: The geometry of the Earth's orbital and rotational motion. The seasons are boreal, i.e., refer to the Northern Hemisphere.

impressive of climatic phenomena, the change of the seasons. The tilt is about 23°, but also this value is subject to slow change.

In describing the movements of the planets, we use a so-called *inertial reference system*, in which Newton's laws of motion apply in their simplest form. The official system, the International Celestial Reference System or Frame[1] (ICRS/ICRF), uses as its origin the common centre of mass of the whole solar system[2]. In many practical applications one rather uses, however, a variant of the reference system in which the origin has been shifted to the centre of mass of the Earth. In this way, one obtains a so-called geocentric, quasi-inertial system. The system is called quasi-inertial because the centre of mass of the Earth is in a permanent state of free fall toward the Sun.

Independent of the location of the origin, all inertial systems have in common that *they are rotation-free*. To realize this using measurements is not trivial; historically, observing the rotation of the Earth was done at astronomical observatories using meridian circles, which were used to register the precise time at which so-called *fundamental stars* would transit the meridian. However, the determination of the places of stars from inside the atmosphere is of limited precision. The astrometric satellite Hipparcos launched in 1989 produced already a major improvement on this, by observing extremely precisely the places of over a hundred thousand stars.

The currently used method for continuously monitoring the rotation of the Earth is VLBI or very long baseline interferometry, which uses radio telescopes around the world. The method's precision is based on the stability of the places of the radio sources or quasars used. These are much further away than the stars we see in the night sky — millions of light years as compared to tens or hundreds of light years — making this a sensible assumption.

It is, however, a matter of philosophical interest that the rotation of the Earth would be the same independent of whether its observation happens by (radio) astronomical means or by using, e.g., a Foucault pendulum or a precise gyroscope, i.e., a closed system. This sameness was already a source of wonderment for Ernst Mach[3].

---

[1]Technically a "reference frame" is the *realization* of a "reference system." In the case of the ICRF, it is a list of precise co-ordinates of celestial objects used in the realization. See IERS, ICRS Centre.

[2]Ephemeris generated in this way, i.e., predictions of the places of the planets, can be found from the web page of the Jet Propulsion Laboratory, JPL, Solar System Dynamics.

[3]Ernst Waldfried Josef Wenzel Mach (1838 – 1916) was an Austrian physicist and philosopher, known for his research into supersonic flow. The dimensionless number used to characterize the velocities of aerial vehicles is named after him.

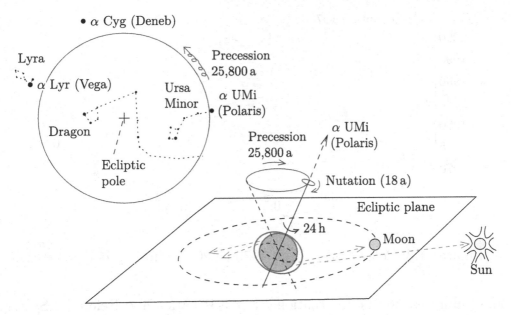

Figure 11.2: Precession, nutation and the torques exerted by the Sun and Moon on the Earth's equatorial bulge.

11.1 Orientation of the Earth

The orientation of the rotation axis of the Earth changes very slowly over time. Relative to the sky, i.e., in an inertial reference system, this motion consists of two parts: *precession* and *nutation*. The cause of the motion is the attraction of Sun and Moon upon the equatorial bulge of the flattened Earth, see Figure 11.2. As both Sun and Moon move in the plane of the ecliptic (Sun) or close to it (Moon), this force, or torque, tries to "put the Earth's axis straight," perpendicular to the ecliptic plane.

If we study changes in the Earth's axis of rotation relative to a reference system attached to the body of the solid Earth, and in the amount of rotational motion, we obtain differently defined quantities:

o Polar motion: this consists of an annual (forced) component, and of a component with a fourteen-month period called the *Chandler[4] wobble*. The amplitude of the motion is several tenths of a second of arc, corresponding to several metres on the Earth's surface, to several metres.

o Length of day (LoD).

Together, all these quantities go by the name of *Earth Orientation Parameters* (EOP). Their changes over time are continuously monitored, and measured

---

[4]Seth Carlo Chandler (1846 – 1913) was an American astronomer.

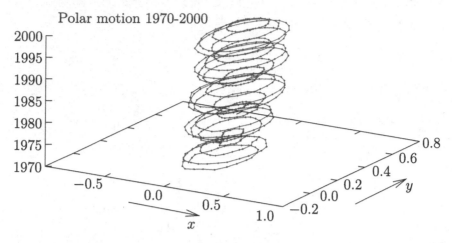

Figure 11.3: Polar motion from 1970 to 2000. Unit: seconds of arc.

values published, by the International Earth Rotation and Reference Systems Service (IERS) on the Internet. The geophysical causes of the temporal changes in these parameters are well understood and can be modelled; one of the main causes of the Chandler wobble are the changes in pressure in the world's oceans and atmosphere (Gross, 2000).

11.2 Polar motion

The direction of the axis of rotation of the Earth relative to the body of the solid Earth varies slightly over time. This *polar motion* has two components, x_p and y_p, the difference of the instantaneous pole from the conventional international orgin (CIO) pole. The first component is in the direction of the Greenwich meridian, and the second is perpendicular to it towards the West, i.e., in the direction of the meridian 90° W. The transformation between the instantaneous and the conventional reference system is done as follows:

$$\overline{\mathbf{X}}_{\text{IT}} = \mathbf{R}_Y\left(x_p\right)\mathbf{R}_X\left(y_p\right)\overline{\mathbf{X}}_{\text{CT}}. \tag{11.1}$$

Here, the matrix $\mathbf{R}_Y$ describes the rotation x_p *around the Y axis*, i.e., the Y axis remains in place, when again the co-ordinates X and Z change. Similarly the matrix $\mathbf{R}_X$ describes the rotation y_p around the X axis, which changes only the co-ordinates Y and Z. The matrices are

$$\mathbf{R}_Y\left(x_p\right) = \begin{bmatrix} \cos x_p & 0 & -\sin x_p \\ 0 & 1 & 0 \\ \sin x_p & 0 & \cos x_p \end{bmatrix} \text{ and } \mathbf{R}_X\left(y_p\right) = \begin{bmatrix} 1 & 0 & 0 \\ 0 & \cos y_p & \sin y_p \\ 0 & -\sin y_p & \cos y_p \end{bmatrix}.$$

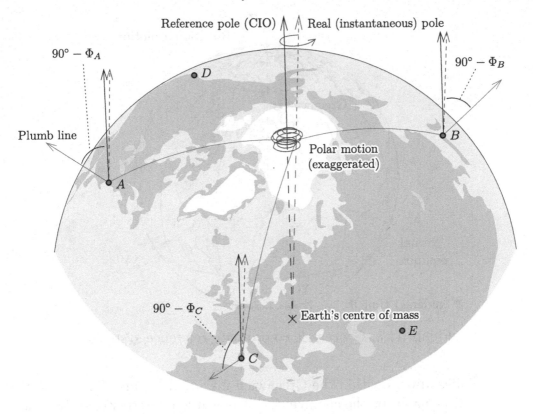

Figure 11.4: How the polar motion has been traditionally observed using six so-called latitude observatories around the globe. A – Gaithersburg MD, USA; B – Mizusawa, Japan; C – Carloforte, Sardinia, Italy; D – Ukiah CA, USA; E – Kitab, Uzbekistan; F – Cincinnati, OH, USA. Nowadays, VLBI is used for this.

Because the angles x_p and y_p are very small, of order seconds of arc, we may approximate $\sin x_\mathrm{p} \approx x_\mathrm{p}$ and $\cos x_\mathrm{p} \approx 1$ (the same for y_p), and also $x_\mathrm{p} y_\mathrm{p} \approx 0$, yielding

$$\mathbf{R}_Y(x_\mathrm{p})\mathbf{R}_X(y_\mathrm{p}) \approx \begin{bmatrix} 1 & 0 & -x_\mathrm{p} \\ 0 & 1 & 0 \\ x_\mathrm{p} & 0 & 1 \end{bmatrix} \begin{bmatrix} 1 & 0 & 0 \\ 0 & 1 & y_\mathrm{p} \\ 0 & -y_\mathrm{p} & 1 \end{bmatrix} = \begin{bmatrix} 1 & 0 & -x_\mathrm{p} \\ 0 & 1 & y_\mathrm{p} \\ x_\mathrm{p} & -y_\mathrm{p} & 1 \end{bmatrix}. \quad (11.2)$$

11.3 Geocentric systems

Figure 11.5 presents a summary of both terrestrial and inertial geocentric co-ordinate reference systems.

XYZ describes an *ECEF* (Earth-Centred, Earth-Fixed) reference system, which rotates along with the body of the solid Earth. Therefore, its X

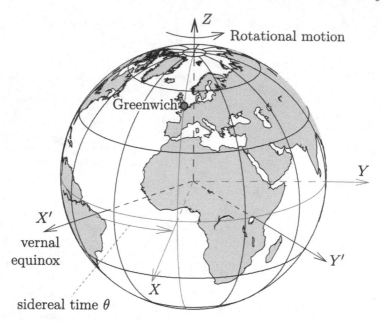

Figure 11.5: Geocentric co-ordinate reference systems.

axis lies always within the Greenwich meridional plane. This system is also designated by the acronym CT (Conventional Terrestrial). Locations on the Earth's surface are nearly fixed, and one can produce maps of them. However, Newton's laws of motion do not work in their basic form.

$X'Y'Z$ describes a (quasi-)*inertial* system, which does not rotate at all (or extremely slowly) with respect to the fixed stars. In such a system, the co-ordinates of the stars will be nearly constant and they may be published. Also, Newton's laws of motion, which describe, e.g., the motions of satellites or gyroscopes, work as such, without terms representing pseudo-forces.

11.3.1   Geocentric terrestrial systems

The geocentric co-ordinate reference systems used by geodesists, like WGS84, are generally defined as follows:

1. The origin of the co-ordinates is in the centre of mass of the Earth.

2. The Z axis of the co-ordinate frame is oriented in the direction of the Earth's axis of rotation, i.e., the North Pole.

3. The X axis of the frame has its direction in the plane of the Greenwich meridian.

Such co-rotating systems are called *terrestrial*, also ECEF. They thus are *not* inertial, and Newton's laws of motion do not apply in their basic form, but they need to be augmented with so-called pseudo-forces like the Coriolis force. When we describe, e.g., the motions of air or water masses on the Earth's surface, this force must necessarily be taken into account. It is the main reason why air and sea currents try to bend to the right on the Northern Hemisphere and to the left on the Southern Hemisphere, a phenomenon that is well known from weather maps[5].

Greenwich was chosen as the world's zero meridian in 1884, in the United States capital of Washington DC, at the International Meridian Conference. Thus was a global standardization achieved not only of longitudes, but also of *time zones*: every zone is fifteen degrees and one hour wide. This was for the young republic, expanding to the West, an existential matter: a conventionally standardized time zone system ensured that trains could run according to a unified timetable.

11.3.2 The conventional terrestrial system

The definition has to be *made more precise* in the following way:

1. Use, as the direction of the Z axis, the so-called CIO, being the mean place of the pole over the years $1900 - 1905$, based on measurements.

   The momentaneous or true pole circles around CIO in a quasi-periodical way, the already mentioned *polar motion*.

2. Today, the plane of the zero meridian is no longer based on observations made at the Greenwich observatory, but rather on global VLBI observations. These are co-ordinated by the International Earth Rotation and Reference Systems Service (IERS). The zero meridian thus is no longer *precisely* the meridian of the Greenwich observatory, though its direction has not changed.

In this way one may create a reference system that co-rotates with the solid Earth, an ECEF reference system, e.g., WGS84 or ITRS*yy* (*yy* being the year number). Another name is Conventional Terrestrial System (CTS).

---

[5]On the other hand, the urban legend that the spinning direction of the water draining from a bathtub would be dictated by the Earth's rotation, is *not* true! A bathtub is simply too small.

11.3.3   Instantaneous terrestrial system

If we take, instead of the conventional pole, the *instantaneous* pole, i.e., the direction of the Earth's axis of rotation, we obtain, instead of the conventional, the so-called *instantaneous terrestrial system* (ITS).

This system must be used, e.g., in connection with astronomical or satellite observations, because it describes the true orientation of the Earth with respect to its own axis of rotation — the instantaneous axis of rotation is the only direction the reference systems of Earth and heavens have in common, the Z axis of both.

Even though also this is a co-rotating system, in this system, unlike in the conventional one and as a result of polar motion, the co-ordinates of points on the Earth's surface will vary a little over time.

The transformation between the conventional and instantaneous systems happens using the *polar motion parameters*: for the already earlier defined x_p, y_p, we obtain with Equations 11.1 and 11.2 given on page 146:

$$
\begin{bmatrix} X \\ Y \\ Z \end{bmatrix}_{IT} = \mathbf{R}_Y(x_p)\,\mathbf{R}_X(y_p) \begin{bmatrix} X \\ Y \\ Z \end{bmatrix}_{CT} \approx \begin{bmatrix} 1 & 0 & -x_p \\ 0 & 1 & y_p \\ x_p & -y_p & 1 \end{bmatrix} \begin{bmatrix} X \\ Y \\ Z \end{bmatrix}_{CT} ,
$$

as the angles x_p, y_p are extremely small.

11.3.4   The geocentric quasi-inertial system

This system is also called the *celestial* or *real astronomical* system (RAS). In Figure 11.5 it is shown with the symbols X' and Y'. Like also the terrestrial systems, this is a *geocentric* system, but it is by its nature inertial or celestial, and the places of the stars are close to fixed in this system. The definition of the system is as follows:

1. The origin of the co-ordinates is again located in the Earth's centre of mass.

2. The Z axis is again oriented along the direction of the Earth's rotation axis, i.e., pointing to the North Pole.

3. *But* the X axis points to the *vernal equinox* in the sky.

Such a reference system does not co-rotate with the solid Earth. It is (to good approximation) *inertial*. We also speak of an *equatorial* co-ordinate system. In this system, the directional co-ordinates are the astronomical right ascension and declination α, δ.

If the co-ordinates of a star on the celestial sphere, α, δ, are known, we may compute the unit direction vector from the observation site to the star as follows:

$$\begin{bmatrix} X \\ Y \\ Z \end{bmatrix}_{\text{RA}} = \begin{bmatrix} \cos\delta\cos\alpha \\ \cos\delta\sin\alpha \\ \sin\delta \end{bmatrix}.$$

Here it should however be considered that, while δ is given in degrees, minutes and seconds, α is given in *units of time*. They must first be converted to degrees, minutes of arc and seconds of arc. One hour corresponds to 15 degrees, one minute to 15 minutes of arc, and one second to 15 seconds of arc.

Going from rectangular co-ordinates to spherical co-ordinates can be done using the following equations. Here we take into use the *half-angle formula*, which yields precise values over the whole value interval and eliminates the irritating problem of finding the right quadrant:

$$\delta = \arcsin Z,$$

$$\alpha = 2\arctan\frac{Y}{X + \sqrt{X^2 + Y^2}}.$$

Again, negative values of α are made positive by adding 24 h.

The co-ordinates of the stars to be used, or *places*, are *apparent*, which as a technical term means "how they appear at a certain point in time[6]." The places α, δ read from a star chart are not apparent. They are places for the co-ordinate frame and situation at an agreed-upon point in time, e.g., 1950.0 or 2000.0. Obtaining apparent places requires a long reduction chain, taking into account precession, nutation, the change in the direction of light caused by the motion of the Earth, i.e., the *aberration*, and also the proper motion of the star and its annual parallax.

The apparent places of stars can be found from the reference work *Apparent Places of Fundamental Stars* precomputed and tabulated by date[7].

The transformation between this, the RAS, and the instantaneous terrestrial or ITS system is

$$\begin{bmatrix} X \\ Y \\ Z \end{bmatrix}_{\text{RA}} = \mathbf{R}_Z(\theta_0)\begin{bmatrix} X \\ Y \\ Z \end{bmatrix}_{\text{IT}} = \begin{bmatrix} \cos(\theta_0) & -\sin(\theta_0) & 0 \\ \sin(\theta_0) & \cos(\theta_0) & 0 \\ 0 & 0 & 1 \end{bmatrix}\begin{bmatrix} X \\ Y \\ Z \end{bmatrix}_{\text{IT}}.$$

Here θ_0 designates the Greenwich Apparent Sidereal Time, or GAST.

---

[6] ...as seen from the centre of the Earth. The fixed stars are however so far away, that the place of the observer does not matter.

[7] The work can nowadays also be read on the Internet: ZAH, Apparent Places.

11.4 Sidereal time

The transformation from the quasi-inertial system to the terrestrial one is done using *sidereal time*.

○ GAST, symbol θ_0.

○ the apparent sidereal time of the observation site, LAST, symbol θ.

$$\theta = \theta_0 + \Lambda^{\mathrm{hms}},$$

in which Λ is the *astronomical longitude* (East) of the observation site, suitably converted to time units.

GAST is

○ the transformation angle between the inertial and terrestrial ("co-rotating") systems, i.e.,

○ the angle describing the orientation of the Earth in an inertial system, i.e.,

○ the difference in longitude or right ascension between the Greenwich meridian and the vernal equinox.

Also GAST is being tabulated, after the fact, when the precise rotational motion of the Earth is known. GAST can be calculated down to about a second accuracy based on calendar and civil time.

If a few minutes accuracy is enough[8], GAST may even be tabulated ready as a function of the day of the year. The table of the annual part of sidereal time per month looks like this:

Month	θ_{m}	Month	θ_{m}	Month	θ_{m}	Month	θ_{m}
Jan.	6 40	April	12 40	July	18 40	Oct.	0 40
Feb.	8 40	May	14 40	Aug.	20 40	Nov.	2 40
March	10 40	June	16 40	Sept.	22 40	Dec.	4 40

In constructing the table, the following information was used: on March 21 at 12 o'clock UTC in Greenwich, the hour angle of the Sun, i.e., of the vernal equinox, i.e., sidereal time, is 0^{h}. This Greenwich sidereal time θ_0 consists of *two parts*: an annual part θ_{a} and clock time t (UTC or Greenwich Mean Time).

---

[8]And if not: ZAH, Apparent Places.

So, the annual part is obtained by subtraction: $\theta_a = \theta_0 - t = -12^h$, i.e., 12^h after adding a full turn or 24^h.

If the sidereal time of March 21, or actually the midnight after the evening of March 20, is $12^h\,00^m$, then the sidereal time of March 0 is $12^h\,00^m - 4 \times 20^m = 10^h\,40^m$. Remember that one day corresponds to about four minutes.

The table is filled using the rule $1\,\text{month} \equiv 2\,\text{hours}$. (It would be possible to make the table a little more precise by considering the varying lengths of the months. However, the cycle of leap years causes an error that is roughly just as big.)

After this, local sidereal time is obtained as follows:

$$\theta = \theta_m + \theta_d + t + \Lambda^{hms}$$
$$= \theta_a + t + \Lambda^{hms} =$$
$$= \theta_0 + \Lambda^{hms}$$

in which

θ_m is the value for the month from the above table,

θ_d is the part for the days within the month, four minutes for every day,

$\theta_a = \theta_m + \theta_d$ is the annual part of sidereal time,

t is the time (UTC[9]), and

Λ is the longitude of the ground station, converted to hours, minutes and seconds ($15° = 1^h$, $1° = 4^m$, $1' = 4^s$).

In Figure 11.6, the following abbreviations have been used:

GAST is Greenwich Apparent Sidereal Time ($= \theta_0$),

LMST is Local Mean Sidereal Time ($= \overline{\theta}$),

h is the hour angle,

h_{Gr} is the hour angle reckoned from Greenwich,

α is the right ascension (of a celestial body), and

Λ is the astronomical longitude (of an object on the Earth's surface).

---

[9]Strictly speaking, UT1, the orientation of the Earth with respect to the mean Sun. The difference is kept to less than a second.

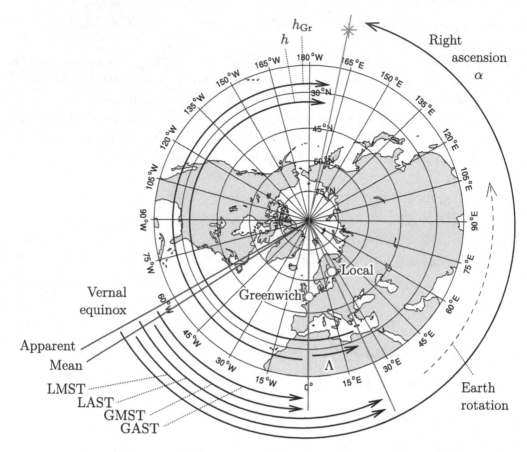

Figure 11.6: Sidereal time.

The place in the sky of the vernal equinox varies in a complicated way due to precession and nutation. Therefore, a distinction must be made between "mean" and "apparent" concepts. The name of the difference is *equation of equinoxes, ee.*

The following equations apply:

$$h = \theta - \alpha,$$
$$h_{Gr} = \theta_0 - \alpha,$$
$$\theta = \theta_0 + \Lambda,$$
$$\overline{\theta} = \overline{\theta}_0 + \Lambda;$$
$$\theta - \overline{\theta} = \theta_0 - \overline{\theta}_0 = ee.$$

11.5 Celestial trigonometry

On the celestial sphere of the observation site, there are at least two different sets of co-ordinates: local co-ordinates and equatorial co-ordinates.

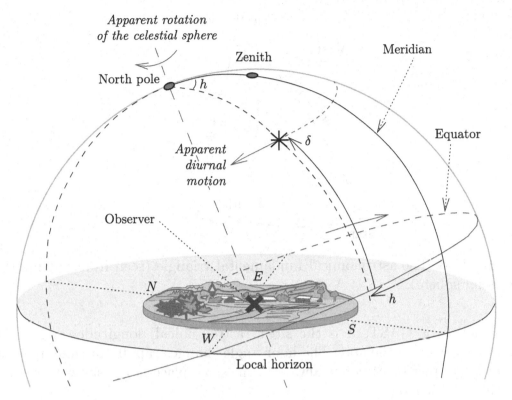

Figure 11.7: The hour angle h and other celestial co-ordinates.

Local spherical co-ordinates are related to local rectangular co-ordinates as follows (x to the North; z to the zenith; also NEU, or North, East, Up):

$$\begin{bmatrix} x \\ y \\ z \end{bmatrix} = \begin{bmatrix} N \\ E \\ U \end{bmatrix} = \begin{bmatrix} \cos A \sin \zeta \\ \sin A \sin \zeta \\ \cos \zeta \end{bmatrix} = \begin{bmatrix} \cos A \cos \eta \\ \sin A \cos \eta \\ \sin \eta \end{bmatrix},$$

in which A is the azimuth (clockwise from the North), ζ is the zenith angle, and $\eta = 90° - \zeta$ is the height or elevation angle. The celestial co-ordinates are again α, δ, right ascension and declination. Their advantage is that the co-ordinates, or "places," of stars in this system are almost constant. Their disadvantage, however, is that the connection with local co-ordinates changes rapidly with the rotational motion of the Earth.

Fortunately, there is an *intermediate form* of co-ordinates: the *hour angle* and declination, h, δ. The hour angle is defined according to Figures 11.7 and 11.8, the angular distance around the Earth's rotation axis (or celestial pole) reckoned from the local meridian of the site.

Equation:

$$h = \theta_0 + \Lambda - \alpha,$$

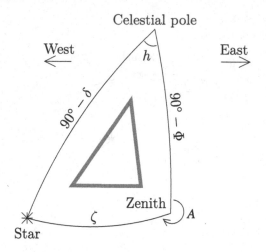

Figure 11.8: The astronomical fundamental triangle (seen from outside the celestial sphere).

in which θ_0 is GAST, Λ is the site's (astronomical) longitude, and α the right ascension of the star. The hour angle changes rapidly in time, but the co-ordinate pair (h, δ) points always to the same place in the sky with respect to the local horizon.

If the star is in the meridian, then $h = 0$ and $\alpha = \theta_0 + \Lambda$. On this is based the use of the transit circle: if, of the three quantities θ_0, α and Λ, two are known, the third may be computed. Depending on the application, we speak of astronomical positioning, determination of time, or determinations of the places of stars. As the saying goes, "One man's noise is another man's signal."

On the celestial sphere, there is an *astronomical fundamental triangle*, Figure 11.8, which consists of the star, the Northern celestial pole, and the zenith. Of the angles of the triangle, we mention the hour angle h (North Pole) and the azimuth A (zenith), of its sides, $90° - \Phi$ (pole-zenith), $90° - \delta$ (star-pole) and zenith angle ζ (star-zenith).

The sine rule:

$$\frac{-\sin A}{\cos \delta} = \frac{\sin h}{\sin \zeta}.$$

The cosine rule:

$$\cos \zeta = \sin \delta \sin \Phi + \cos \delta \cos \Phi \cos h,$$
$$\sin \delta = \sin \Phi \cos \zeta + \cos \Phi \sin \zeta \cos A.$$

Compute first, using the cosine rule, either δ or ζ, and then, using the sine rule, either h or A. Thus, one obtains the conversion in both directions: $(A, \zeta) \longleftrightarrow (h, \delta)$.

11.6 Rotation matrices and celestial co-ordinate systems

The transformations of the various celestial co-ordinate systems may also be derived in rectangular co-ordinates using rotation matrices.

Let the vector (the length of which we assume 1, i.e., it is only a direction vector)

$$\mathbf{x} = x\mathbf{i} + y\mathbf{j} + z\mathbf{k} = X\mathbf{i}' + Y\mathbf{j}' + Z\mathbf{k}'.$$

In this, $\{\mathbf{i}, \mathbf{j}, \mathbf{k}\}$ is a topocentrically oriented orthonormal basis, $\{\mathbf{i}', \mathbf{j}', \mathbf{k}'\}$ a basis that is oriented according to the local astronomical system, i.e., it is equatorial.

In spherical co-ordinates, we have

$$\begin{bmatrix} x \\ y \\ z \end{bmatrix} = \begin{bmatrix} \cos A \sin \zeta \\ \sin A \sin \zeta \\ \cos \zeta \end{bmatrix} \quad \text{and} \quad \begin{bmatrix} X \\ Y \\ Z \end{bmatrix} = \begin{bmatrix} \cos \alpha \cos \delta \\ \sin \alpha \cos \delta \\ \sin \delta \end{bmatrix}.$$

Transform as follows:

1. The direction of the x axis is changed from North to South by doing an inversion with the matrix $\mathbf{M}_1$, which changes the topocentric co-ordinate system from left-handed to right-handed.

2. The new pair of axes xz is rotated by an angle of $90° - \Phi$ to the North, i.e., in the negative direction with respect to the y axis, where Φ is the astronomical latitude. In this way, the z axis will be parallel to the Z axis of the geocentric system, i.e., the rotation axis of the Earth.

3. The new pair of axes xy is rotated by an angle of θ around the z axis in the negative direction, i.e., to the West, and clockwise from the local meridian to the vernal equinox. The angle θ is the local (apparent) sidereal time.

The matrices are all orthogonal:

$$\mathbf{M}_1 = \begin{bmatrix} -1 & 0 & 0 \\ 0 & 1 & 0 \\ 0 & 0 & 1 \end{bmatrix},$$

$$\mathbf{R}_2 = \begin{bmatrix} \cos(90° - \Phi) & 0 & \sin(90° - \Phi) \\ 0 & 1 & 0 \\ -\sin(90° - \Phi) & 0 & \cos(90° - \Phi) \end{bmatrix} = \begin{bmatrix} \sin \Phi & 0 & \cos \Phi \\ 0 & 1 & 0 \\ -\cos \Phi & 0 & \sin \Phi \end{bmatrix},$$

$$\mathbf{R}_3 = \begin{bmatrix} \cos \theta & -\sin \theta & 0 \\ \sin \theta & \cos \theta & 0 \\ 0 & 0 & 1 \end{bmatrix}.$$

Compute

$$\mathbf{R_3 R_2 M_1} = \begin{bmatrix} \cos\theta & -\sin\theta & 0 \\ \sin\theta & \cos\theta & 0 \\ 0 & 0 & 1 \end{bmatrix} \begin{bmatrix} -\sin\Phi & 0 & \cos\Phi \\ 0 & 1 & 0 \\ \cos\Phi & 0 & \sin\Phi \end{bmatrix} =$$

$$= \begin{bmatrix} -\cos\theta\sin\Phi & -\sin\theta & \cos\theta\cos\Phi \\ -\sin\theta\sin\Phi & \cos\theta & \sin\theta\cos\Phi \\ \cos\Phi & 0 & \sin\Phi \end{bmatrix}.$$

After this:

$$\begin{bmatrix} \cos\alpha\cos\delta \\ \sin\alpha\cos\delta \\ \sin\delta \end{bmatrix} = \begin{bmatrix} -\cos\theta\sin\Phi & -\sin\theta & \cos\theta\cos\Phi \\ -\sin\theta\sin\Phi & \cos\theta & \sin\theta\cos\Phi \\ \cos\Phi & 0 & \sin\Phi \end{bmatrix} \begin{bmatrix} \cos A\sin\zeta \\ \sin A\sin\zeta \\ \cos\zeta \end{bmatrix},$$

from which we immediately identify

$$\sin\delta = \sin\Phi\cos\zeta + \cos\Phi\sin\zeta\cos A,$$

which is the cosine rule in the fundamental triangle star–celestial pole–zenith.

The inverse transformation is, because of the orthogonality of the transformation matrix (transpose!),

$$\begin{bmatrix} \cos A\sin\zeta \\ \sin A\sin\zeta \\ \cos\zeta \end{bmatrix} = \begin{bmatrix} -\cos\theta\sin\Phi & -\sin\theta\sin\Phi & \cos\Phi \\ -\sin\theta & \cos\theta & 0 \\ \cos\theta\cos\Phi & \sin\theta\cos\Phi & \sin\Phi \end{bmatrix} \begin{bmatrix} \cos\alpha\cos\delta \\ \sin\alpha\cos\delta \\ \sin\delta \end{bmatrix}.$$

With these, the transformation of the spherical co-ordinates may be done though the three-dimensional direction cosines.

Exercises

Exercise 11 – 1:   Co-ordinate transformation on the celestial sphere

Let the rectangular co-ordinates of a star in the equatorial system be

$$\overline{\mathbf{X}} = \begin{bmatrix} X \\ Y \\ Z \end{bmatrix} = \begin{bmatrix} \cos\alpha\cos\delta \\ \sin\alpha\cos\delta \\ \sin\delta \end{bmatrix},$$

and in the ecliptical system

$$\overline{\mathbf{X}}' = \begin{bmatrix} X' \\ Y' \\ Z' \end{bmatrix} = \begin{bmatrix} \cos\ell\cos b \\ \sin\ell\cos b \\ \sin b \end{bmatrix},$$

in which ℓ, b are ecliptical longitude and latitude.

Between the equator and the ecliptic, there is an angle $\epsilon = 23°26'$ (epoch J2000.0). Both the right ascension α and the ecliptical longitude ℓ are reckoned positive in the counter-clockwise direction, i.e., going to the East, and the declination δ as well as the ecliptical latitude b are reckoned positive to the North.

a. If

$$\overline{X}' = R\overline{X},$$

find the matrix R.

b. If

$$\overline{X} = S\overline{X}',$$

what is the matrix S?

c. The star Vega (α Lyrae) is at the place $\alpha = 18^h36^m56^s$, $\delta = 38°47'$ (epoch J2000.0). Calculate ℓ, b.

d. Derive closed formulas using spherical trigonometry, see Chapter 7. Calculate anew ℓ, b. Verify your solution.

Chapter 12

The orbital motion of satellites

12.1 Kepler's laws

The idea to launch artificial satellites is old. The notion started to become a reality in the 1950s with the development of rocket technology. As part of the International Geophysical Year (IGY) in 1957–1958, the United States planned to launch a small satellite for geophysical research, using a new, three-stage Vanguard rocket. Technical problems however delayed the launch, and to everybody's surprise on October 4, 1957, the Soviet Union succeeded, as the first state, in injecting a satellite, the Sputnik 1, into orbit around the Earth.

On-board Sputnik was only a battery driven radio beacon, which soon fell silent. Radio amateurs were among the first to observe the new celestial body. Also hobbyists with an interest in space technology made visual observations of Sputnik and of the even brighter end stage of the launcher, which had also reached orbit. Using these valuable observations, it was possible to track Sputnik's rapidly changing orbit (Auchincloss, 1957).

Today, thousands of satellites orbit the Earth, most of them very small. However, a few hundred of them — generally the end stages of launchers — are so large as to be visible to the naked eye, after dark in the light of the Sun. With binoculars, they are very easily observed. The orbital heights of these easily observable satellites range from 400 kilometres to over one thousand kilometres. The inclinations vary over a large range, but certain inclination values, like 56°, 65°, 72°, 74°, 81°, 90° and 98° curry particular favour[1].

Then there are the *Iridium satellites*, which form a group of their own. These have remained in space after a commercially failed mobile telephony project. On each Iridium satellite there is a long, specularly reflective metal antenna, which casts the light of the Sun extremely brightly to the observer, whenever the geometry is right[2].

---

[1]Perhaps this is due to safety-of-habitation constraints around the launch site, as SeeSat-L, Visual Satellite Observing FAQ, explains.

[2]Predictions tailored to your observation location may be obtained from the website Heavens Above. It appears, however, that in the future, the Iridium constellation will no longer produce predictable flares.

Figure 12.1: Tycho Brahe observing using his huge wall quadrant.

Satellites are artificial celestial bodies and move along their orbits following the same laws that were first elucidated by Kepler for the planets of our solar system.

Johannes Kepler (1571–1630) was a German astronomer, mathematician, mystic and *avant-la-lettre* nerd (Koestler, 1968, 1960). As the assistant of the Danish nobleman Tycho Brahe, he got his hands on Brahe's unique observational material on the planet Mars.

Tycho Brahe (1546–1601) was a colourful individual with a short temper — he had a silver partial nose prosthesis replacing the piece he had lost in a duel. He was, however, patient and painstaking as an observer.

Brahe built an observatory on the island of Hven, in the middle of the Danish Sound, given to him by the Danish King Frederik II in 1576. He collected untiringly — with the naked eye, using a huge wall quadrant — a uniquely precise and extensive observational material on Mars.

Up until Brahe's time, it was customary to model the motions of the planets as circles — for practical applications such as the drawing of horoscopes — by combining larger and smaller circles, so-called epicycles, into a complicated mathematical machinery. Kepler succeeded in making this traditional, circles based model match Brahe's observations to within some eight minutes of arc. Any other astronomer back then would have been happy with this, but not Kepler. He knew, as also Brahe himself, that the material was better than that, a precision of order one minute of arc. For comparison, consider that the disc of Jupiter and the crescent of Venus are at best also of size one minute of arc, and each can be seen as a non-pointlike object only in binoculars.

Kepler attempted, using pen and paper, a careful triangulation of the orbit of Mars. He used measurements made at different times, when Mars was in its orbit at the same place, but the Earth at a different place — and, conversely, using measurements when the Earth was at the same place but Mars at a different place. The task, the reconstruction of the orbits of both the Earth and Mars around the Sun, was not an easy one.

The method is presented in Figures 12.2 and 12.3. The first figure shows, how one can build, from observations of many Mars oppositions, a timetable which tells in which direction Mars is seen — in other words, the ecliptic longitude of Mars — by an observer on the Sun at any calendar day. It is here assumed that the orbits of Earth and Mars lie in the same plane. This is not precisely true because the inclination of the Mars orbit is almost two degrees[3]. Furthermore, it is of course assumed that the "timetable" metaphor is valid and that Mars really returns to the same place in space after completing a full orbit, i.e., one period. According to what we know now, this does not hold precisely.

The second figure tells us how one may carry out a triangulation. From the observations one obtains directly the direction to Mars, the geocentric ecliptic longitude as seen from Earth. From the date of the observation again one obtains, according to the method shown in Figure 12.2, also the *heliocentric* ecliptic longitude. Thus one may, by means of triangulation[4], compute the true place of Mars in its orbital plane.

The final outcome was that the orbit could not be a circle. It was clearly eccentric — the eccentricity of the Mars orbit is 0.09. Only the eccentricity of Mercury's orbit is larger. Mercury is so close to the Sun that its observation was difficult before the age of the telescope. The value of the eccentricity means that the place of the Sun has shifted from the centre of the Mars orbit to the side, by an amount of 9% of the radius of the orbit. But that alone was not enough. Somehow the orbit was distorted, and the place of Mars in the sky was often several minutes of arc wrong. It took a while to grasp that the orbit was an *ellipse*.

---

[3]In fact, Kepler developed his own method for the reduction of observations of Mars to the plane of the Earth's orbit, the ecliptic, which is not at all simple.

[4]In geodesy, the method is called *intersection*. In astronomy, it is also referred to as *annual parallax*, typically used for determining distances to other stars. Even the nearest stars have parallaxes of less than one second of arc.

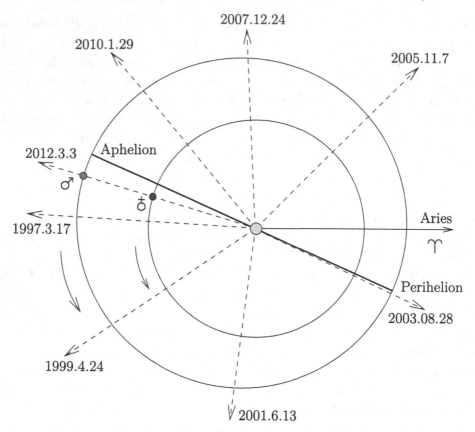

Figure 12.2: A "direction timetable" for the planet Mars may be constructed using oppositions. The straight line is the major axis of the Mars orbit, connecting perihelion and aphelion, the points closest to and farthest from the Sun. "Aries" is the vernal equinox, i.e., the zero direction on the ecliptic.

In the case of Mars, the difference between a circle and an ellipse was very small; in Figures 12.2 and 12.3, it would not be much more than the thickness of the drawn orbit curve. It was only the unusually large eccentricity of the Mars orbit together with Tycho Brahe's exceptional observational skill, and Johannes Kepler's perseverance, which in the end solved this riddle — a science story that has deservedly inspired poetry (Williams, 1868).

After the initial realization, Kepler quickly found the solution to the whole enigma of the orbits and motions of the planets. Kepler's laws are[5]:

I The orbits of the planets are *ellipses*, each in its own orbital plane, and the Sun is located in one of the focal points of the ellipse,

---

[5]Of the laws, the third was found ten years after the first and second ones. It was published in the work *Harmonices mundi* (1619) whereas the second and first laws appeared in the work *Astronomia nova* (1609). The second law was found before the first one.

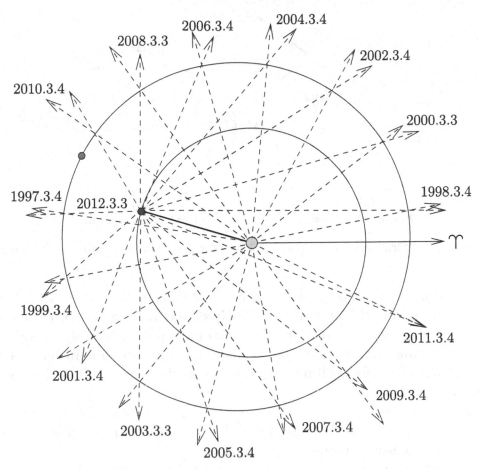

Figure 12.3: The precise figure of the orbit of the planet Mars can be determined by triangulation, by using observations made on the same day of every year. The line connecting Sun and Earth is the base used for triangulation.

II the velocity of orbital motion varies in such a way that in a unit of time, the radius between Sun and planet sweeps always over the same amount of surface area ("the law of areas," i.e., conservation of angular momentum[6]), and

III the squares of the *periods* of the various planets stand in the same proportion as the cubes of the mean radii, or semi-major axes, of their orbital ellipses.

In the case of satellites, however, it is the centre of mass of the Earth, not the Sun, that is the centre of orbital motion and the main source of the force shaping the orbit.

---

[6]Kepler already had a hunch that the orbital motion of the planets was caused by some influence emanating from the Sun.

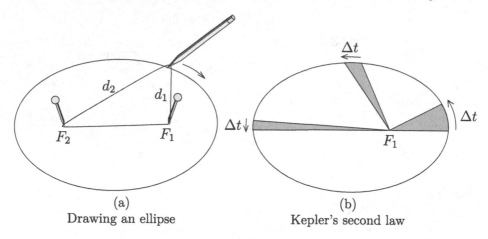

(a)
Drawing an ellipse

(b)
Kepler's second law

Figure 12.4: Drawing an ellipse, and Kepler's second law of orbital motion.

Let us shortly discuss relevant properties of ellipses. See Figure 12.4. When drawing an ellipse, the property is exploited that it is the set of points, the sum of the distances of which from the two focal points, $d_1 + d_2$, is constant. This also means that if the ellipse is made from reflective material, a lamp placed in one of the foci will project an image in the other focus. This explains their name.

Johannes Kepler's second law of orbital motion, the law of areas, says that in the same amount of time Δt the radius between Sun and planet sweeps always over the same amount of surface area.

In advanced theory, the state of motion of the satellite is described using six so-called Kepler orbital elements, which may (slowly) evolve over time. See Figure 12.5 and Table 12.1. In Table 12.2 on page 169 again we see in light of examples, how the height of a satellite above the Earth's surface determines the satellite's orbital period P.

The life of Johannes Kepler was not easy. His father was a mercenary, who presumably was killed in the Dutch war of independence. His mother practiced herbal medicine. Still at an advanced age she was charged with witchcraft. Johannes found himself having to provide for his mother's defence, and succeeded in freeing her in the end. Kepler fell ill and died at age 58 in 1630 in Regensburg, Bavaria, where he was buried. As part of the ongoing thirty-year war, Swedish troops — which might well have included Finns — destroyed the graveyard. Kepler's mortal remains have never been found.

Table 12.1: Kepler's orbital elements.

Angular elements describing the orbital *orientation* in space:

Ω right ascension or astronomical longitude of the ascending node. The zero point for this longitude is the place on the celestial sphere where the plane of the zodiac, or ecliptic, and the equatorial plane intersect, and where the Sun crosses the equator at the beginning of spring, i.e., the vernal equinox.

i inclination or tilt of the orbital plane with respect to the equator. For GPS satellites, $i = 55°$. This is also the highest latitude (North and South) where the satellite moves through the zenith.

ω argument of perigee. The angular distance between the ascending node and the perigee of the satellite orbit.

Elements describing *size and shape* of the orbit:

a semi-major axis of the satellite orbit, or the mean radius of the orbit. The mean height of the satellite is obtained by subtracting from this the radius of the Earth.

e eccentricity of the satellite orbit. $1 - e^2 = \frac{b^2}{a^2}$, in which b is the semi-minor axis. This describes how much the centre of the Earth is offset from the centre of the orbital ellipse, i.e., how large is the height difference between the perigee (lowest point of the orbit) and apogee (highest point).

The element describing the satellite's *place in its orbit*, the timetable (three alternatives):

$M(t)$ mean anomaly: $M(t) \stackrel{\text{def}}{=} \frac{2\pi(t-\tau)}{P}$, in which τ is the time of perigee passage, and P is the period.

$E(t)$ eccentric anomaly: $E(t) = M(t) + e \sin E(t)$.

$\nu(t)$ true anomaly: $\dfrac{\tan \frac{1}{2}\nu(t)}{\tan \frac{1}{2}E(t)} = \sqrt{\dfrac{1+e}{1-e}}$.

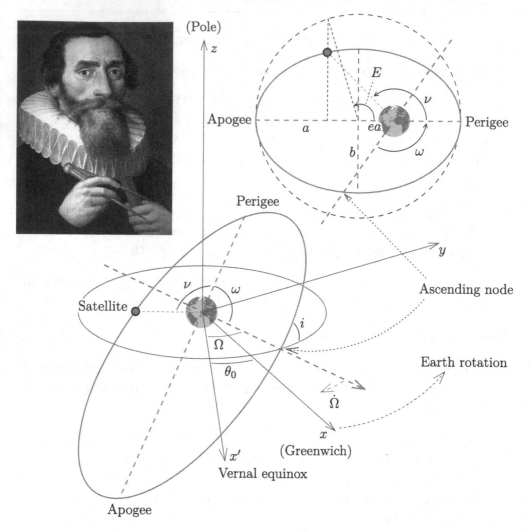

Figure 12.5: Kepler orbital elements for an Earth orbiting satellite.

12.2   The circular orbit approximation in manual computation

Currently many good quality sources of predictions of satellite orbits for visual or radio observations can be found, e.g., the excellent software gpredict. Its computation is based on professional level algorithms and it is able to download satellite ephemeris from the Internet. Also the earlier mentioned site, Heavens Above, offers good quality predictions containing information important to visual observers, like the brightness or *magnitude*, and the satellite's path plotted on a star chart.

In the following, we describe a much simpler problem: computing a circular satellite orbit around a spherical Earth. This is a sufficiently good method to

Table 12.2: Kepler's third law: there exists a simple functional relationship between a satellite's orbital height and period. These are selected, illustrative values.

Height (km)	Period	Remark
0	$84^{\mathrm{m}}29^{\mathrm{s}}$	Schuler[†] period
400	$92^{\mathrm{m}}34^{\mathrm{s}}$	
800	$100^{\mathrm{m}}52^{\mathrm{s}}$	
20 183	$11^{\mathrm{h}}58^{\mathrm{m}}$	GPS
35 785	$23^{\mathrm{h}}56^{\mathrm{m}}$	Geostationary
376 603	$27^{\mathrm{d}}07^{\mathrm{h}}$	Moon

[†]Maximilian Schuler (1882 – 1972) was a German navigation technologist.

make visual satellite observations possible, and to find a satellite even without using computer technology.

In the case of a circular orbit, the eccentricity e is zero, and the argument of perigee ω is without meaning, because the perigee and apogee are no longer defined. Also all three anomalies, ν, E and M, are now identical. Therefore, only four orbital elements are left:

o the right ascension of the ascending node Ω,

o the orbital inclination i,

o the semi-major axis of the orbit, which now is simply the radius of the circular orbit, and

o the orbital "anomaly" $\nu = M = E$, now reckoned from the ascending node.

When we know the time t_0 of the satellite's equator crossing, i.e., the orbit's *ascending node*, and the right ascension (inertial or celestial longitude) Ω, we may compute the corrections to time and longitude for various latitudes, as described below.

12.3 Crossing a certain latitude in the inertial system

In the following we assume the Earth to be a sphere.

If the target latitude ϕ is given, we may compute the distance ν from the ascending node (this distance is sometimes called the "downrange angle") in angular units as follows:

$$\sin \nu = \frac{\sin \phi}{\sin i},$$

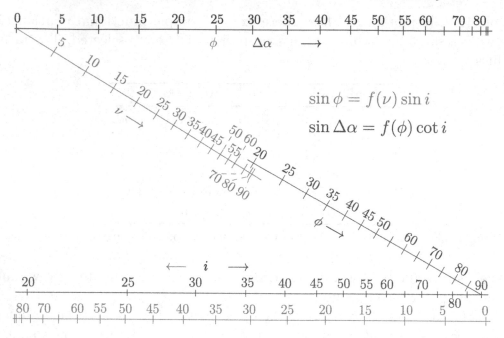

$$\sin \phi = f(\nu) \sin i$$

$$\sin \Delta\alpha = f(\phi) \cot i$$

Figure 12.6: Nomogram for determining ν and $\Delta\alpha$ from inclination i and latitude ϕ.

in which i is the inclination of the satellite orbit. See Figure 12.6. From this again follows the elapsed time with Kepler's third law. The period is

$$P = \sqrt{\frac{4\pi^2}{GM}a^3}.$$

From this

$$\Delta t = \frac{\nu}{2\pi}P,$$

the time of flight from the equator to latitude ϕ. The azimuth angle of the satellite track with the local meridian is obtained from Clairaut's equation (Section 8.2 on page 93):

$$\cos \phi \sin A = \cos (0) \cos i,$$

because on the equator $(\phi = 0)$ $A = \dfrac{\pi}{2} - i$, so

$$\sin A = \frac{\cos i}{\cos \phi}.$$

Now the difference in right ascension with the equator crossing, using the sine rule for a spherical triangle, is

$$\frac{\sin \Delta\alpha}{\sin A} = \frac{\sin \phi}{\sin i} \implies \sin \Delta\alpha = \frac{\tan \phi}{\tan i} = \tan \phi \cot i.$$

After this, we obtain the satellite's right ascension and time when crossing latitude circle ϕ:

$$t = t_0 + \Delta t,$$
$$\alpha = \Omega + \Delta\alpha.$$

Here, Ω is the *right ascension of the ascending node* of the satellite orbit.

Both longitude and right ascension α (and $\Delta\alpha$) are reckoned positive to the East.

12.4 Topocentric co-ordinates of the satellite

Usually the *height* of the satellite orbit h is given. According to its definition, the distance of the satellite from the Earth's centre is

$$r = \|\mathbf{r}\| = a_e + h,$$

in which a_e is the equatorial radius of the Earth. In spherical approximation $a_e = R$. Then, the rectangular, inertial co-ordinates of the satellite are

$$\mathbf{r} = \begin{bmatrix} r\cos\phi\cos\alpha_G \\ r\cos\phi\sin\alpha_G \\ r\sin\phi \end{bmatrix},$$

in which ϕ is the satellite's geocentric latitude (i.e., the geocentric declination δ_G) and α_G the geocentric right ascension. See Figure 12.7.

Let the geocentric latitude of the ground station be Φ and its longitude Λ. At the moment t, the *geocentric right ascension of the ground station* is

$$\theta = \Lambda + \theta_0 = \Lambda + t + \theta_a,$$

in which t is universal time and θ_a the annual part of sidereal time. In this, θ equals local sidereal time, which thus is *the orientation of the local meridian* in (inertial) space.

Now the rectangular, inertial co-ordinates of the ground station are

$$\mathbf{R} = \begin{bmatrix} R\cos\Phi\cos\theta \\ R\cos\Phi\sin\theta \\ R\sin\Phi \end{bmatrix}.$$

Subtraction yields the difference vector:

$$\mathbf{r} - \mathbf{R} \stackrel{\text{def}}{=} \mathbf{d} = \begin{bmatrix} d_1 \\ d_2 \\ d_3 \end{bmatrix} \begin{bmatrix} d\cos\delta_T\cos\alpha_T \\ d\cos\delta_T\sin\alpha_T \\ d\sin\delta_T \end{bmatrix},$$

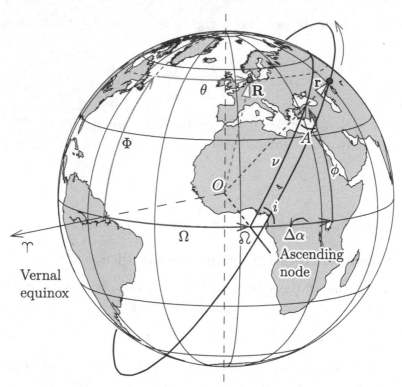

Figure 12.7: Places of the satellite ($\mathbf{r}$) and the ground station ($\mathbf{R}$) in the inertial system.

from which we obtain as the solution *topocentric spherical co-ordinates*:

$$\tan \delta_{\mathrm{T}} = \frac{d_3}{\sqrt{d_1^2 + d_2^2}}, \qquad \tan \alpha_{\mathrm{T}} = \frac{d_2}{d_1}.$$

These are thus the right ascension and the declination of the satellite as seen from the observation site against the background of the stars. With a star chart and binoculars, we can now make observations of the satellite.

We also obtain the *distance* of the satellite:

$$d = \|\mathbf{d}\| = \sqrt{d_1^2 + d_2^2 + d_3^2}.$$

Using the distance, one may compute the visual brightness or *magnitude* of the satellite. The distances are generally of class $500 - 1000\,\mathrm{km}$, and the magnitudes $2 - 5$.

12.5 Latitude crossing in the terrestrial system

If the terrestrial system is used in the computations, the *longitude* of the satellite's equator crossing, $\lambda_0 = \Omega - \theta_0(t_0)$, must be known, in which θ_0 is GAST at the moment of equator crossing t_0.

The time difference Δt from the equator crossing to latitude Φ is obtained in the same way. Now, however, we compute the *longitude difference* as follows:

$$\Delta\lambda = \Delta\alpha - \Delta t \cdot \omega,$$

in which ω is the *angular rate of the Earth's rotation*, about $0°.25$ per minute. Thus is obtained the longitude of the satellite:

$$\lambda = \lambda_0 + \Delta\lambda.$$

The geocentric co-ordinates of both the ground station and the satellite may also be described in the *terrestrial system*. In this system, the co-ordinates of ground station and satellite are

$$\mathbf{R} = R \begin{bmatrix} \cos\Phi\cos\Lambda \\ \cos\Phi\sin\Lambda \\ \sin\Phi \end{bmatrix}, \mathbf{r} = r \begin{bmatrix} \cos\phi\cos\lambda \\ \cos\phi\sin\lambda \\ \sin\phi \end{bmatrix}.$$

From this again is obtained the *terrestrial* topocentric vector:

$$\mathbf{r} - \mathbf{R} \stackrel{\text{def}}{=} \mathbf{d} = \begin{bmatrix} d\cos\delta_T\cos(\Lambda - h_T) \\ d\cos\delta_T\sin(\Lambda - h_T) \\ d\sin\delta_T \end{bmatrix},$$

from which may be solved the topocentric declination δ_T and *hour angle* h_T.

If we then want to determine the right ascension to be used with a star chart, we only need to subtract this hour angle from local sidereal time:

$$\alpha_T = \theta - h_T = (\theta_0 + \Lambda) - h_T = \theta_0 + (\Lambda - h_T).$$

Here θ_0 is the Greenwich sidereal time and $(\Lambda - h_T)$ the hour angle from Greenwich.

12.6 Determining the orbit from observations

See Figure 12.8. It suffices for an approximate orbit determination that the place of the satellite in the sky has been observed on two different moments in time t_1 and t_2 — i.e., given are $(\alpha(t_1), \delta(t_1))$ and $(\alpha(t_2), \delta(t_2))$ — and of course the times t_1 and t_2. Such observations do not require any high technology. Many hobbyists have made these already since the time of Sputnik: binoculars, a star chart, and a couple of stopwatches are sufficient. Of course, also a short-wave radio receiver is needed for the time signals[7]. The method is today only of historical interest, but we shall describe it here.

First, we compute the topocentric direction vectors (unit vectors):

$$\mathbf{e}(t_1) = \frac{\mathbf{d}(t_1)}{d(t_1)} = \begin{bmatrix} \cos \delta(t_1) \cos \alpha(t_1) \\ \cos \delta(t_1) \sin \alpha(t_1) \\ \sin \delta(t_1) \end{bmatrix},$$

$$\mathbf{e}(t_2) = \frac{\mathbf{d}(t_2)}{d(t_2)} = \begin{bmatrix} \cos \delta(t_2) \cos \alpha(t_2) \\ \cos \delta(t_2) \sin \alpha(t_2) \\ \sin \delta(t_2) \end{bmatrix}.$$

When also the ground-station vector $\mathbf{R}$ has been computed, we may compute $d(t_1), d(t_2)$ using the cosine rule, if a suitable value[8] for the satellite height h — or, equivalently, for the radius of the satellite orbit $r = R + h$ — has been given:

$$r^2 = R^2 + d^2 + 2Rd \cos(\angle \mathbf{e}, \mathbf{R}) \implies d^2 + 2Rd \sin \eta + R^2 - r^2 = 0$$

$$\implies d_{1,2} = \frac{-2R \sin \eta \pm \sqrt{4R^2 \sin^2 \eta - 4(R^2 - r^2)}}{2} =$$

$$= -R \sin \eta \pm \sqrt{r^2 - R^2(1 - \sin^2 \eta)},$$

in which

$$\sin \eta = \cos(\angle \mathbf{e}, \mathbf{R}) = \langle \mathbf{e} \cdot \mathbf{e}_U \rangle,$$
$$(\text{i.e., } \sin \eta(t_i) = \cos(\angle \mathbf{e}(t_i), \mathbf{R}) = \langle \mathbf{e}(t_i) \cdot \mathbf{e}_U \rangle, \quad i = 1, 2)$$

is the projection of the satellite's direction vector on the local plumb line, and

$$\mathbf{e}_U = \frac{\mathbf{R}}{R} = \begin{bmatrix} \cos \Phi \cos \theta \\ \cos \Phi \sin \theta \\ \sin \Phi \end{bmatrix},$$

is the zenith vector of the observer, i.e., the unit vector pointing straight up. The value $\sin \eta$, the sine of the elevation angle, is directly computable as the dot product $\langle \mathbf{e} \cdot \mathbf{e}_U \rangle$, when both vectors have been given in rectangular components.

---

[7]Nowadays a precise and reliable source for official time is the Internet service NTP, *Network Time Protocol*. A wireless alternative is the long-wave radio station DCF77 in Frankfurt am Main, Germany, which, on a carrier frequency of 77.5 kHz, transmits a signal that synchronizes special clocks all over Europe.

[8]An experienced observer can estimate the height of the satellite from its speed of motion in the sky with surprising precision.

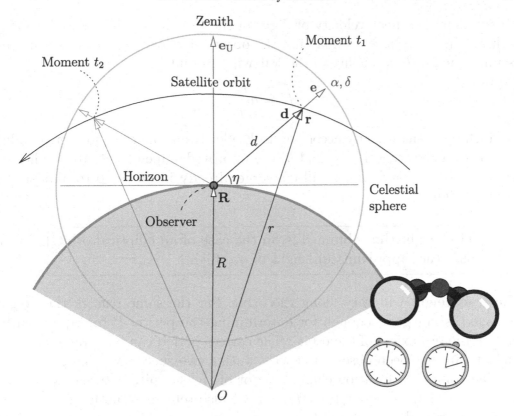

Figure 12.8: The geometry of satellite orbit determination.

Only the positive solution makes physical sense:

$$d = \sqrt{r^2 - R^2 \left(1 - \sin^2 \eta\right)} - R \sin \eta.$$

Thus is obtained $d(t_1), d(t_2)$ and thus $\mathbf{d}(t_1) = d(t_1)\mathbf{e}(t_1), \mathbf{d}(t_2) = d(t_2)\mathbf{e}(t_2)$.
For the satellite velocity is obtained

$$v = \frac{\|\mathbf{d}(t_2) - \mathbf{d}(t_1)\|}{t_2 - t_1}.$$

We know, however, what the velocity for a circular orbit *should* be at height h:
according to Kepler's third law

$$P = \sqrt{\frac{4\pi^2}{GM} \left(R + h\right)^3} \implies v_{\mathrm{K}} = \frac{2\pi \left(R + h\right)}{P} = \sqrt{\frac{GM}{R + h}}.$$

We precompute the following table:

Height h (km)	500	750	1000	1500
Velocity v_{K} ($\mathrm{m/s}$)	7612.609	7477.921	7350.139	7113.071

We see that the linear velocity of flight of the satellite diminishes only slowly with height. Therefore, we may use the observed velocity v for correcting the assumed height h according to the following formula:

$$h' = h \frac{v_K}{v},$$

in which v_K is the velocity according to Kepler (from the table for height value h), v the calculated velocity, and h' the improved value of the satellite height. Iteration with this formula will converge already in one step to almost the correct height.

> The height thus obtained is, in the case of an elliptical orbit, only the (approximate) height of overflight!

The true height will vary along the orbit. For the same reason, the height obtained is not good enough for determining the period P (or equivalently: the semi-major axis a of the orbit). The correct period can be obtained only, if the satellite has been observed at least during two successive nights.

The computed location change vector of the satellite between two observation epochs $\mathbf{d}(t_2) - \mathbf{d}(t_1) = \mathbf{r}(t_2) - \mathbf{r}(t_1)$ also tells us something about the satellite's orbital inclination. Compute the cross-product (vectorial product) of the location change vector with, e.g., the satellite's geocentric location vector $\mathbf{r}(t_1) = \mathbf{R} + \mathbf{d}(t_1)$, as follows:

$$(\mathbf{d}(t_2) - \mathbf{d}(t_1)) \times \mathbf{r}(t_1) = \|\mathbf{d}(t_2) - \mathbf{d}(t_1)\| \, \|\mathbf{r}(t_1)\| \begin{bmatrix} \sin i \sin \Omega \\ -\sin i \cos \Omega \\ \cos i \end{bmatrix},$$

from which i may be solved — the vector in square brackets is actually the unit vector perpendicular to the orbital plane, called $\mathbf{s}$ in Figure 12.9. Also Ω and (with P) t_0 may now be computed, see Figure 12.9. Now we may already start generating predictions!

When the height of the satellite (and her approximate period) as well as the inclination are known, one may compute also the fast precessional motion of the ascending node (the formula applies for a circular orbit):

$$\Omega(t) = \Omega(t_0) + (t - t_0) \frac{d\Omega}{dt} =$$

$$= \Omega(t_0) - (t - t_0) \cdot \frac{3}{2} \sqrt{\frac{GM}{a^3}} \left(\frac{a_e}{a}\right)^2 J_2 \cos i,$$

in which a is the semi-major axis of the satellite orbit, a_e the equatorial radius of the Earth ellipsoid, and J_2 the Earth's *dynamic flattening*, value

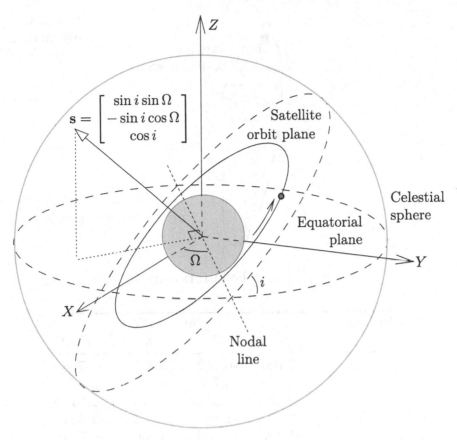

Figure 12.9: Determination of inclination i and right ascension of the ascending node Ω. The vector $\mathbf{s}$ is perpendicular to the plane of the satellite orbit.

$J_2 = 1082.62 \cdot 10^{-6}$. One of the first achievements of the satellite era was the precise determination of J_2[9]. As a numerical formula:

$$\frac{d\Omega}{dt} = -6.529\,27 \cdot 10^{24} \frac{\cos i}{a^{3.5}} \quad \left[\mathrm{m}^{3.5\ \mathrm{degrees}}/\mathrm{day}\right]$$

(note the unit!) Thus the following table is obtained (in units of degrees per day):

---

[9]In fact, this motion of the ascending node is so large, that without taking it into account it is impossible to generate sensible orbit predictions.

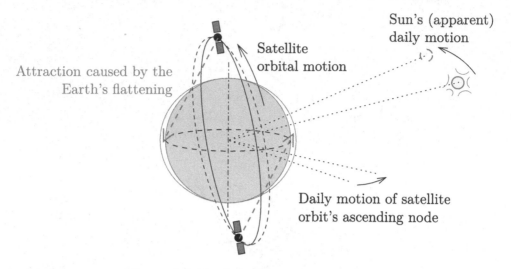

Figure 12.10: A heliostationary orbit.

Inclination	Height (km)			
(°)	500	750	1000	1500
0	-7.651	-6.752	-5.985	-4.758
56	-4.278	-3.776	-3.347	-2.661
65	-3.233	-2.854	-2.529	-2.011
74	-2.109	-1.861	-1.650	-1.311
81	-1.197	-1.056	-0.936	-0.744
90	0.0	0.0	0.0	0.0
0.9856	97°.401	98°.394	99°.478	101°.955

The last row of the table is different. The value $0.9856\,^{\text{degrees}}/\text{day}$ is the *apparent angular rate of the Sun* against the background of the stars. If we tune the precession of the satellite's ascending node, i.e., the precession of the orbital plane, to this value, the satellite will fly always over the same area at the same local solar time. In this way we achieve a *heliostationary orbit*, also — if we are at a sufficient height — a "no-shadow" orbit, the benefit of which is continuous light on the solar panels, and in remote sensing, always the same solar elevation angle when imaging the surface of the Earth. See Figure 12.10. Given in the table is the inclination angle which yields, for every average height in the table's column, just such a Sun-stationary orbit.

The precession rate of the orbital plane of the satellite *relative to the Sun* is now

$$q \stackrel{\text{def}}{=} \frac{d\Omega}{dt} - 0.9856\,^{\text{degrees}}/\text{day}.$$

For example, for a satellite at a height of $500\,\text{km}$ and with an orbital inclination of $56°$:

$$q = (-4.278 - 0.9856)\,^{\text{degrees}}/\text{day} = 5.2636\,^{\text{degrees}}/\text{day},$$

and the period of one turn of the orbital plane is

$$\frac{360}{5.2636} \text{ days} = 68.4 \text{ days}.$$

This is roughly the time that elapses until the satellite becomes again visible in the evening sky of the same latitude zone. As this amounts to several months, it is not likely that the satellite will be found on the predicted place on its orbit, if the orbital period has not been determined very precisely. This succeeds only, if the satellite has been observed at least twice during its previous period of visibility, and if the effect of air drag may be ignored; e.g., at a height of 500 km one may not do so. In this case, the only possibility to find the satellite again is by monitoring the predicted orbit in the hope that the satellite will show up in the field of view of the binoculars independently of the calculated timetable. In the worst case, one may end up waiting one and a half hours.

Exercises

Exercise 12 – 1:
The mean height of a satellite orbit (circular) is 500 km. Calculate the period.

Exercise 12 – 2:
The inclination angle of a satellite orbit is 65°. Calculate the motion of the ascending node $\frac{d\Omega}{dt}$, in units of degrees per day.

Exercise 12 – 3:
On April 9, in the evening at 19:15 UTC, a satellite crosses the equator flying North. The right ascension of the ascending node is 135°. Calculate

a. at what moment in time the satellite flies over latitude 60°, and

b. at which right ascension.

Exercise 12 – 4:
The latitude of Helsinki is $\Phi = 60°$ and its longitude $\Lambda = 25°$. Calculate the geocentric α, δ for the moment given in the previous exercise.

Chapter 13

The surface theory of Gauss

Johann Carl Friedrich Gauss (1777−1855), often referred to as *Princeps mathematicorum* ("the foremost of mathematicians"), was among the first mathematicians to consider non-Euclidean geometry, like his contemporaries Bolyai[1] and Lobachevsky[2].

See Figure 13.1. Gauss was also among the first to develop a theory of curved surfaces. His theory is still based on the assumption that the two-dimensional surface is embedded in a three-dimensional space. The derivations and operations derived are in that case simple. Nevertheless, the theory can be applied as such to the investigation of the curved surface of the Earth. Gauss was also a geodesist, who measured and computed the geodetic networks of both Hannover and Braunschweig using the method of least squares.

Gauss was a universal genius. He was extremely productive, but in many ways peculiar. Some suspect that he — like so many other great scientists — was afflicted with the As-

Figure 13.1: Portrait of Carl Friedrich Gauss aged 50. Lithography, Siegfried Detlev Bendixen 1828.

perger syndrome. He was, e.g., very reluctant to publish anything that was in the least unfinished. His motto was *pauca sed matura* ("little, but mature"). We may only guess how many mathematical inventions (including the fast Fourier transform) had to be re-invented, decades or even more than a century later, because of this reluctance.

Gauss has also appeared in literary fiction (Kehlmann, 2006).

---

[1]János Bolyai (1802−1860) was a Hungarian mathematician and a pioneer of non-Euclidean geometry.

[2]Nikolai Ivanovich Lobachevsky (1792−1856) was a Russian mathematician and a pioneer of hyperbolic geometry.

Let a curved surface S be given in a three-dimensional, Euclidean space $\mathbb{R}^3$. The surface is *parametrized* with the parameters (u, v); for example, the surface of the Earth, parametrization (φ, λ).

In three-dimensional space we may create, in order to decribe the location of a point, an orthonormal basis

$$\{\mathbf{i}, \mathbf{j}, \mathbf{k}\}.$$

On this basis, a point, or a location vector, is[3]

$$\mathbf{x} = x^1\mathbf{i} + x^2\mathbf{j} + x^3\mathbf{k} = x\mathbf{i} + y\mathbf{j} + z\mathbf{k}.$$

Often we write for $\mathbf{x}$ its *representation* on this basis (and, sloppily, leave off the overbar where the context is clear):

$$\overline{\mathbf{x}} = \begin{bmatrix} x^1 \\ x^2 \\ x^3 \end{bmatrix} = \begin{bmatrix} x \\ y \\ z \end{bmatrix}.$$

These three parameters thus form a parametrization of the three-dimensional space $\mathbb{R}^3$.

13.1   A curve in space

A curve C running through space can be parametrized by a parameter s. Then, the points on the curve are $\mathbf{x}(s)$ for different values of s. If it holds for the parameter that

$$ds^2 = dx^2 + dy^2 + dz^2,$$

we say that C has been *parametrized by distance* (or arc length).

Examples.

1. The numbers on a measuring tape form a parametrization by distance.

2. When driving along a road, the numbers on the trip meter form a parametrization by distance of the road.

3. On Mannerheim Road in Helsinki, the co-ordinate x (Northing) in the Finnish National Map Grid Co-ordinate System (KKJ) forms a parametrization, though *not* by distance (because the direction of the road, while generally North-South, is varying).

---

[3]The superscripts in this formula are not powers! We shall return to this notation later on.

Let us assume in the sequel, that the parametrization used is unambiguous and differentiable[4] (and thus continuous).

The *tangent vector* of curve C is obtained by differentiation:

$$\mathbf{t}(s) = \frac{d\mathbf{x}(s)}{ds} = \mathbf{x}_s;$$

the length of the tangent is

$$\|\mathbf{t}\| = \sqrt{\left(\frac{dx}{ds}\right)^2 + \left(\frac{dy}{ds}\right)^2 + \left(\frac{dz}{ds}\right)^2} = \sqrt{\frac{dx^2 + dy^2 + dz^2}{ds^2}} = \sqrt{1} = 1.$$

This *only* holds if s is a parametrization by distance.

An arbitrary parametrization t may always be converted to a parametrization by distance in the following way:

$$s(t) = \int_0^t \frac{ds}{d\tau} d\tau = \int_0^t \frac{\sqrt{dx^2 + dy^2 + dz^2}}{d\tau} d\tau = \qquad (13.1)$$

$$= \int_0^t \sqrt{\left(\frac{dx}{d\tau}\right)^2 + \left(\frac{dy}{d\tau}\right)^2 + \left(\frac{dz}{d\tau}\right)^2} d\tau,$$

or in differential form

$$ds = dt \sqrt{\left(\frac{dx(t)}{dt}\right)^2 + \left(\frac{dy(t)}{dt}\right)^2 + \left(\frac{dz(t)}{dt}\right)^2};$$

thus, $s(t)$ may always be computed by integrating 13.1.

Differentiating once more produces the *curvature vector* of the curve:

$$\mathbf{k}(s) = \frac{d\mathbf{t}(s)}{ds} = \frac{d^2}{ds^2} \mathbf{x}(s).$$

See Figure 13.2, where we have drawn also the *osculating circle*[5]. This is the circle, the plane of which contains both the tangent and the curvature vector, and of which both tangent and curvature are locally the same as those of curve C. We may also think that the osculating circle has two successive tangent vectors in common with curve C: $\mathbf{t}(t)$ and $\mathbf{t}(t + dt)$, where we then let dt go to the limit $dt \to 0$.

---

[4] All of its partial derivatives exist and they are continuous.
[5] Latin for "kissing circle." The name was given by Leibniz.

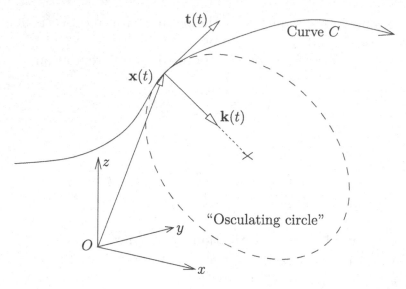

Figure 13.2: Tangent vector $\mathbf{t}$, curvature vector $\mathbf{k}$, and osculating circle of curve C.

13.2   The first fundamental form (metric)

The *first fundamental form* according to Gauss is:

$$ds^2 = E\,du^2 + 2F\,dudv + G\,dv^2. \tag{13.2}$$

(u, v) form the *parametrization* of the surface, ds is a *distance element* inside the surface.

We will see later that this fundamental form is the same as the *metric g_{ij}* of the surface in question, and an alternative notation for this is

$$ds^2 = g_{11}du^2 + g_{12}dudv + g_{21}dvdu + g_{22}dv^2.$$

If the point $\mathbf{x}$ lies on the surface S, we may expand its *derivative* into the derivatives of its components or co-ordinates:

$$\mathbf{x}_u = \frac{\partial \mathbf{x}}{\partial u} = \frac{\partial x}{\partial u}\mathbf{i} + \frac{\partial y}{\partial u}\mathbf{j} + \frac{\partial z}{\partial u}\mathbf{k},$$

$$\mathbf{x}_v = \frac{\partial \mathbf{x}}{\partial v} = \frac{\partial x}{\partial v}\mathbf{i} + \frac{\partial y}{\partial v}\mathbf{j} + \frac{\partial z}{\partial v}\mathbf{k}.$$

Or, as abstract component vectors:

$$\overline{\mathbf{X}}_u = \begin{bmatrix} \partial x/\partial u \\ \partial y/\partial u \\ \partial z/\partial u \end{bmatrix},\ \overline{\mathbf{X}}_v = \begin{bmatrix} \partial x/\partial v \\ \partial y/\partial v \\ \partial z/\partial v \end{bmatrix}.$$

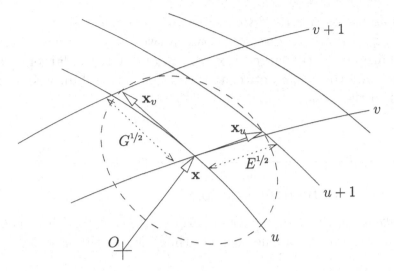

Figure 13.3: The first fundamental form of Gauss.

We call these vectors the *tangent vectors* of surface S and parametrization (u, v).

From these we obtain, as the scalar or dot products of two vectors, the *elements of the fundamental form*:

$$E = \langle \mathbf{x}_u \cdot \mathbf{x}_u \rangle, \quad F = \langle \mathbf{x}_u \cdot \mathbf{x}_v \rangle, \quad G = \langle \mathbf{x}_v \cdot \mathbf{x}_v \rangle.$$

In Figure 13.3 the first fundamental form of Gauss and the tangent vectors $\mathbf{x}_u, \mathbf{x}_v$ are shown. It is readily shown — chain rule in three dimensions (x, y, z) — that

$$ds^2 = dx^2 + dy^2 + dz^2 =$$

$$= \left[\left(\frac{\partial x}{\partial u} \right)^2 + \left(\frac{\partial y}{\partial u} \right)^2 + \left(\frac{\partial z}{\partial u} \right)^2 \right] du^2 +$$

$$+ 2 \left[\left(\frac{\partial x}{\partial u} \frac{\partial x}{\partial v} \right) + \left(\frac{\partial y}{\partial u} \frac{\partial y}{\partial v} \right) + \left(\frac{\partial z}{\partial u} \frac{\partial z}{\partial v} \right) \right] dudv +$$

$$+ \left[\left(\frac{\partial x}{\partial v} \right)^2 + \left(\frac{\partial y}{\partial v} \right)^2 + \left(\frac{\partial z}{\partial v} \right)^2 \right] dv^2 =$$

$$= \langle \mathbf{x}_u \cdot \mathbf{x}_u \rangle \, du^2 + 2 \langle \mathbf{x}_u \cdot \mathbf{x}_v \rangle \, dudv + \langle \mathbf{x}_v \cdot \mathbf{x}_v \rangle \, dv^2,$$

from which the form 13.2 follows directly.

We can see, e.g., that in the direction of the v curves $(dv = 0)$:

$$ds^2 = E \, du^2,$$

i.e., E represents the *metric distance* between two successive $(u, u+1)$ curves. Similarly G represents the distance between two successive v curves. The more the curves are apart, the larger is $\mathbf{x}_u$ or $\mathbf{x}_v$, and also the larger is E or G. F again represents the *angle* between the u and v curve bundles: it vanishes if the angle is a right one.

13.3   The second fundamental form

The *normal* on a two-dimensional surface is the vector which is orthogonal to every curve running on the surface, thus also to the u and v curves. We write

$$\mathbf{n} = n_1\mathbf{i} + n_2\mathbf{j} + n_2\mathbf{k},$$

or

$$\overline{\mathbf{n}} = \begin{bmatrix} n_1 \\ n_2 \\ n_2 \end{bmatrix}.$$

Then

$$\langle \mathbf{n} \cdot \mathbf{x}_u \rangle = \langle \mathbf{n} \cdot \mathbf{x}_v \rangle = 0. \tag{13.3}$$

We require also that the length of the vector is 1:

$$\|\mathbf{n}\|^2 = n_1^2 + n_2^2 + n_3^2 = 1.$$

We differentiate $\mathbf{x}$ a second time:

$$\mathbf{x}_{uu} = \frac{\partial^2 \mathbf{x}}{\partial u^2}, \qquad \mathbf{x}_{uv} = \frac{\partial \mathbf{x}}{\partial u \partial v}, \qquad \mathbf{x}_{vv} = \frac{\partial \mathbf{x}}{\partial v^2}.$$

Now the second fundamental form of Gauss is

$$e\, du^2 + 2f\, du\, dv + g\, dv^2,$$

in which

$$e = \langle \mathbf{n} \cdot \mathbf{x}_{uu} \rangle, \qquad f = \langle \mathbf{n} \cdot \mathbf{x}_{uv} \rangle, \qquad g = \langle \mathbf{n} \cdot \mathbf{x}_{vv} \rangle. \tag{13.4}$$

Based on condition 13.3 we also have

$$0 = \frac{\partial}{\partial u} \langle \mathbf{n} \cdot \mathbf{x}_u \rangle = \langle \mathbf{n}_u \cdot \mathbf{x}_u \rangle + \langle \mathbf{n} \cdot \mathbf{x}_{uu} \rangle \implies e = -\mathbf{n}_u \cdot \mathbf{x}_u,$$

$$0 = \frac{\partial}{\partial v} \langle \mathbf{n} \cdot \mathbf{x}_u \rangle = \langle \mathbf{n}_v \cdot \mathbf{x}_u \rangle + \langle \mathbf{n} \cdot \mathbf{x}_{uv} \rangle \implies f = -\mathbf{n}_v \cdot \mathbf{x}_u,$$

$$0 = \frac{\partial}{\partial u} \langle \mathbf{n} \cdot \mathbf{x}_v \rangle = \langle \mathbf{n}_u \cdot \mathbf{x}_v \rangle + \langle \mathbf{n} \cdot \mathbf{x}_{uv} \rangle \implies f = -\mathbf{n}_u \cdot \mathbf{x}_v, \tag{13.5}$$

$$0 = \frac{\partial}{\partial v} \langle \mathbf{n} \cdot \mathbf{x}_v \rangle = \langle \mathbf{n}_v \cdot \mathbf{x}_v \rangle + \langle \mathbf{n} \cdot \mathbf{x}_{vv} \rangle \implies g = -\mathbf{n}_v \cdot \mathbf{x}_v.$$

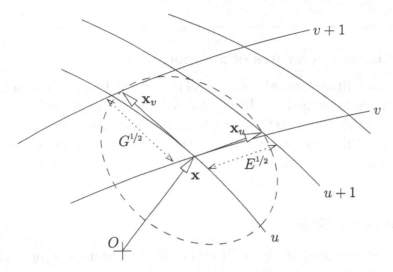

Figure 13.4: A geometric view of the second fundamental form of Gauss.

See Figure 13.4. When moving from location to location, the direction of the normal vector changes: when we move from point $\mathbf{x}$ (u, v) to point $\mathbf{x}'$ $(u + du, v)$, the normal changes $\mathbf{n} \rightarrow \mathbf{n}' = \mathbf{n} + \dfrac{\partial \mathbf{n}}{\partial u} du$. Similarly, when we move from point $\mathbf{x}$ to point $\mathbf{x}''$ $(u, v + dv)$, the normal changes $\mathbf{n} \rightarrow \mathbf{n}'' = \mathbf{n} + \dfrac{\partial \mathbf{n}}{\partial v} dv$. The total change will be

$$d\mathbf{n} = \frac{\partial \mathbf{n}}{\partial u} du + \frac{\partial \mathbf{n}}{\partial v} dv = \mathbf{n}_u \, du + \mathbf{n}_v \, dv.$$

The norm or length of the normal vector is always 1. Therefore, the vector can only be changed in two directions, either that of the tangent vector $\mathbf{x}_u$, or that of the tangent vector $\mathbf{x}_v$.

Let us separate them by *projection*:

$$\langle d\mathbf{n} \cdot \mathbf{x}_u \rangle = \langle \mathbf{n}_u \cdot \mathbf{x}_u \rangle \, du + \langle \mathbf{n}_v \cdot \mathbf{x}_u \rangle \, dv,$$
$$\langle d\mathbf{n} \cdot \mathbf{x}_v \rangle = \langle \mathbf{n}_u \cdot \mathbf{x}_v \rangle \, du + \langle \mathbf{n}_v \cdot \mathbf{x}_v \rangle \, dv,$$

in which we may directly identify the elements of the second fundamental form, e, f, g, using the above formulas 13.5:

$$- \langle d\mathbf{n} \cdot \mathbf{x}_u \rangle = e \, du + f \, dv,$$
$$- \langle d\mathbf{n} \cdot \mathbf{x}_v \rangle = f \, du + g \, dv.$$

Unlike for the first fundamental form, there does not exist anything corresponding to the second fundamental form in Riemann's surface theory. It only exists for surfaces that are *embedded* in a surrounding (Euclidean) space.

Alternatively, there exists a tensorial notation:

$$\beta_{11} = e, \qquad \beta_{12} = \beta_{21} = f, \qquad \beta_{22} = g.$$

13.4   Principal curvatures of a surface

The second fundamental form of Gauss describes the curvature of the surface in space by visualizing how the direction of the normal vector changes when we move in the direction of either the u or the v co-ordinate curves. Unfortunately, this is not enough to describe the curvature in an *absolute* sense, because the parametrization (u, v)

1. is arbitrary, and

2. is not metrically scaled.

The last point means that, if the direction of the normal vector **n** changes by an amount $d\mathbf{n}$ when we travel a distance du along the v co-ordinate curve, we do not yet know *to how many metres du corresponds*. If the distance is long, then a small change $d\mathbf{n}$ of the normal vector means only a small curvature of the surface. If the distance is short, the same change $d\mathbf{n}$ corresponds to a large curvature of the surface.

The parameter curves $u = \text{constant}$ and $v = \text{constant}$ are generally not orthogonal to each other, which complicates the issue further.

Write the first and the second fundamental forms into matrix form:

$$\mathbf{H} = \begin{bmatrix} E & F \\ F & G \end{bmatrix}, \mathbf{B} = \begin{bmatrix} e & f \\ f & g \end{bmatrix}.$$

Form the matrix

$$\mathbf{C} \stackrel{\text{def}}{=} \mathbf{H}^{-1}\mathbf{B} = \frac{1}{EG - F^2} \begin{bmatrix} G & -F \\ -F & E \end{bmatrix} \cdot \begin{bmatrix} e & f \\ f & g \end{bmatrix} =$$

$$= \frac{1}{EG - F^2} \begin{bmatrix} Ge - Ff & Gf - Fg \\ Ef - Fe & Eg - Ff \end{bmatrix}.$$

The matrix **C** is called the *shape operator*, and the above equations are called the Weingarten[6] equations. See Wikipedia, Shape operator.

We may say that the multiplication with the inverse of the **H** matrix, i.e., the first fundamental form describing *the length of a distance element*, performs *a metric scaling* of the **B** matrix[7].

---

[6]Julius Weingarten (1836 – 1910) was a German mathematician.

[7]A more technical description: **B** is a covariant tensor β_{ij}, and **H** the covariant metric tensor g_{ij}. $\mathbf{H}^{-1}$ again corresponds to the contravariant tensor g^{ij}. **C** is now a "mixed tensor" $\beta_k^i = g^{ij}\beta_{jk}$, for which the *tensorial* eigenvalue problem is

$$\left(\beta_j^i - \kappa\delta_j^i\right) x^j = 0.$$

The equation is the same as Equation 13.6. See Subsection 14.3.2 on page 214.

Principal curvatures:

The matrix $\mathbf{C}^8$ has *two eigenvalues*: the values $\kappa_{1,2}$ for which

$$(\mathbf{C} - \kappa\mathbf{I})\,\mathbf{x} = 0 \tag{13.6}$$

for suitable pairs of values $\mathbf{x} = \begin{bmatrix} du & dv \end{bmatrix}^\mathsf{T}$. The solutions $\kappa_{1,2}$ are called the *principal curvatures* of the surface S. They are *invariants* with respect to the chosen parametrization (u, v). The corresponding eigenvectors, the pairs of values $\mathbf{x}_1 = \begin{bmatrix} du_1 & dv_1 \end{bmatrix}^\mathsf{T}$ and $\mathbf{x}_2 = \begin{bmatrix} du_2 & dv_2 \end{bmatrix}^\mathsf{T}$, define the local *principal directions of curvature* on the surface.

Other invariants:

1. The product $\kappa_1\kappa_2 = \det\mathbf{C} = \dfrac{\det\mathbf{B}}{\det\mathbf{H}} = \dfrac{eg - f^2}{EG - F^2}$ is the Gaussian *total curvature*.

2. The half-sum $\frac{1}{2}(\kappa_1 + \kappa_2) = \frac{1}{2}(\mathbf{C}_{11} + \mathbf{C}_{22}) = \frac{1}{2}\dfrac{eG + gE - 2fF}{EG - F^2}$ is Germain[9]'s *mean curvature*.

Principal directions of curvature:

We may also look into the *eigenvectors* of $\mathbf{C}$, which are called principal directions of curvature. To this end, we write

$$0 = \mathbf{H}\,(\mathbf{C} - \kappa_1\mathbf{I})\,\mathbf{x}_1 = (\mathbf{B} - \kappa_1\mathbf{H})\,\mathbf{x}_1,$$
$$0 = \mathbf{H}\,(\mathbf{C} - \kappa_2\mathbf{I})\,\mathbf{x}_2 = (\mathbf{B} - \kappa_2\mathbf{H})\,\mathbf{x}_2.$$

Multiply the first from the left with $\mathbf{x}_2^\mathsf{T}$ and the second with $\mathbf{x}_1^\mathsf{T}$, and take the transpose of the result:

$$0 = \mathbf{x}_2^\mathsf{T}\mathbf{B}\mathbf{x}_1 - \kappa_1\mathbf{x}_2^\mathsf{T}\mathbf{H}\mathbf{x}_1,$$
$$0 = \left(\mathbf{x}_1^\mathsf{T}\mathbf{B}\mathbf{x}_2 - \kappa_2\mathbf{x}_1^\mathsf{T}\mathbf{H}\mathbf{x}_2\right)^\mathsf{T} = \mathbf{x}_2^\mathsf{T}\mathbf{B}\mathbf{x}_1 - \kappa_2\mathbf{x}_2^\mathsf{T}\mathbf{H}\mathbf{x}_1,$$

because both $\mathbf{B}$ and $\mathbf{H}$ are symmetric matrices. We obtain

$$\left(\mathbf{x}_1^\mathsf{T}\mathbf{B}\mathbf{x}_2\right)^\mathsf{T} = \mathbf{x}_2^\mathsf{T}\mathbf{B}\mathbf{x}_1 \quad \text{and} \quad \left(\mathbf{x}_1^\mathsf{T}\mathbf{H}\mathbf{x}_2\right)^\mathsf{T} = \mathbf{x}_2^\mathsf{T}\mathbf{H}\mathbf{x}_1.$$

---

[8] Why $\mathbf{C} = \mathbf{H}^{-1}\mathbf{B}$? Why not $\mathbf{C}^\mathsf{T} = \mathbf{B}\mathbf{H}^{-1}$? We shall see in Chapter 14 Section 14.3.2 on page 214, that these alternatives are only two of the generally four different ways of writing the eigenvalue problem, which all yield the same eigenvalues and essentially the same eigenvectors.

[9] Marie-Sophie Germain (1776–1831) was an autodidactic French mathematical genius doing research into number theory and the mathematics of elasticity, as well as a philosopher. She corresponded on number theory with Gauss, who considered her his peer. Wikipedia, Sophie Germain.

Subtracting the first from the second leads to the equation

$$\left(\kappa_2 - \kappa_1\right) \mathbf{x}_2^{\mathsf{T}} \mathbf{H} \mathbf{x}_1 = 0.$$

This shows that, *if the principal radii of curvatures differ*[10], *then the expression* $\mathbf{x}_2^{\mathsf{T}} \mathbf{H} \mathbf{x}_1$ *vanishes*. This expression[11] can be interpreted *as a dot product*: we may, e.g., write symbolically $\langle \mathbf{x}_2 \cdot \mathbf{x}_1 \rangle = 0$. Actually in rectangular co-ordinates in the tangent plane the matrix $\mathbf{H}$ is the unit matrix $\mathbf{H} = \mathbf{I}$, and furthermore $\mathbf{x}_2^{\mathsf{T}} \mathbf{x}_1 = 0$, i.e., $\mathbf{x_1}$ and $\mathbf{x_2}$ are mutually orthogonal in the Euclidean sense: $\mathbf{x_1} \perp \mathbf{x_2}$.

> The principal directions of curvature are mutually orthogonal.

This is a special case of the general rule, that a self-adjoint operator has mutually orthogonal eigenvectors. As another example of this, we may mention the eigenfunctions of the differential equations of Sturm[12]–Liouville theory. See Wikipedia, Sturm–Liouville theory.

Example. On the surface of an ellipsoidal Earth, the co-ordinates of a point are

$$\mathbf{x} = \begin{bmatrix} N(\varphi) \cos \varphi \cos \lambda \\ N(\varphi) \cos \varphi \sin \lambda \\ N(\varphi) \left(1 - e^2\right) \sin \varphi \end{bmatrix}.$$

From this

$$\mathbf{x}_\varphi = \frac{\partial \mathbf{x}}{\partial \varphi} = \begin{bmatrix} \cos \lambda \dfrac{d}{d\varphi} \left(N(\varphi) \cos \varphi\right) \\ \sin \lambda \dfrac{d}{d\varphi} \left(N(\varphi) \cos \varphi\right) \\ \left(1 - e^2\right) \dfrac{d}{d\varphi} \left(N(\varphi) \sin \varphi\right) \end{bmatrix}.$$

Using equations derived in Appendix C on page 249 we obtain

$$\mathbf{x}_\varphi = M(\varphi) \begin{bmatrix} -\sin \varphi \cos \lambda \\ -\sin \varphi \sin \lambda \\ +\cos \varphi \end{bmatrix}.$$

$$\mathbf{x}_\lambda = \frac{\partial \mathbf{x}}{\partial \lambda} = N(\varphi) \begin{bmatrix} -\cos \varphi \sin \lambda \\ +\cos \varphi \cos \lambda \\ 0 \end{bmatrix}.$$

---

[10]If they are equal, any linear combination of $\mathbf{x}_1$ and $\mathbf{x}_2$ will again be an eigenvector. Two of these can always be chosen to be mutually orthogonal.

[11]In index notation: $g_{ij} x_2^i x_1^j$.

[12]Jacques Charles François Sturm (1803–1855) was a French mathematician.

The surface normal is obtained as a vectorial or cross-product, after normalization:

$$\mathbf{n} = \frac{\langle \mathbf{x}_\varphi \times \mathbf{x}_\lambda \rangle}{\|\langle \mathbf{x}_\varphi \times \mathbf{x}_\lambda \rangle\|},$$

where

$$\langle \mathbf{x}_\varphi \times \mathbf{x}_\lambda \rangle = NM \begin{bmatrix} -\cos^2 \varphi \cos \lambda \\ -\cos^2 \varphi \sin \lambda \\ -\sin \varphi \cos \varphi \cos^2 \lambda - \sin \varphi \cos \varphi \sin^2 \lambda \end{bmatrix} =$$

$$= -NM \begin{bmatrix} \cos^2 \varphi \cos \lambda \\ \cos^2 \varphi \sin \lambda \\ \sin \varphi \cos \varphi \end{bmatrix} = -NM \cos^2 \varphi \begin{bmatrix} \cos \lambda \\ \sin \lambda \\ \tan \varphi \end{bmatrix},$$

the norm of which is

$$\|\langle \mathbf{x}_\varphi \times \mathbf{x}_\lambda \rangle\| = NM \cos^2 \varphi \sqrt{1 + \tan^2 \varphi} = NM \cos \varphi.$$

So

$$\mathbf{n} = - \begin{bmatrix} \cos \varphi \cos \lambda \\ \cos \varphi \sin \lambda \\ \sin \varphi \end{bmatrix}, \tag{13.7}$$

not a surprising result.

Let us calculate the first fundamental form:

$$E = \langle \mathbf{x}_\varphi \cdot \mathbf{x}_\varphi \rangle = M^2 \left(\sin^2 \varphi \left(\sin^2 \lambda + \cos^2 \lambda \right) + \cos^2 \varphi \right) = M^2,$$

$$F = \langle \mathbf{x}_\varphi \cdot \mathbf{x}_\lambda \rangle = 0,$$

$$G = \langle \mathbf{x}_\lambda \cdot \mathbf{x}_\lambda \rangle = N^2 \cos^2 \varphi = p^2.$$

For the second fundamental form, we calculate

$$\mathbf{n}_\varphi = \begin{bmatrix} +\sin \varphi \cos \lambda \\ +\sin \varphi \sin \lambda \\ -\cos \varphi \end{bmatrix}, \quad \mathbf{n}_\lambda = \begin{bmatrix} +\cos \varphi \sin \lambda \\ -\cos \varphi \cos \lambda \\ 0 \end{bmatrix}$$

and thus we obtain (Equations 13.5 on page 186)

$$e = - \langle \mathbf{n}_\varphi \cdot \mathbf{x}_\varphi \rangle = +M,$$

$$f = - \langle \mathbf{n}_\varphi \cdot \mathbf{x}_\lambda \rangle = - \langle \mathbf{n}_\lambda \cdot \mathbf{x}_\varphi \rangle = 0,$$

$$g = - \langle \mathbf{n}_\lambda \cdot \mathbf{x}_\lambda \rangle = +N \cos^2 \varphi.$$

This means

$$\mathbf{H} = \begin{bmatrix} E & F \\ F & G \end{bmatrix} = \begin{bmatrix} M^2 & 0 \\ 0 & N^2 \cos^2 \varphi \end{bmatrix},$$

$$\mathbf{B} = \begin{bmatrix} e & f \\ f & g \end{bmatrix} = \begin{bmatrix} M & 0 \\ 0 & N \cos^2 \varphi \end{bmatrix},$$

$$\mathbf{C} = \mathbf{H}^{-1} \mathbf{B} = \begin{bmatrix} \frac{1}{M} & 0 \\ 0 & \frac{1}{N} \end{bmatrix}.$$

The *principal curvatures* $\kappa_{1,2}$ are the *eigenvalues* of $\mathbf{C}$, solutions of the eigenvalue problem

$$(\mathbf{C} - \kappa \mathbf{I})\,\mathbf{x} = 0;$$

the values are obtained by solving the zero points of the corresponding, so-called *characteristic polynomial*:

$$\det(\mathbf{C} - \kappa \mathbf{I}) = 0 \implies \det \begin{bmatrix} \frac{1}{M} - \kappa & 0 \\ 0 & \frac{1}{N} - \kappa \end{bmatrix} = 0$$

$$\implies \left(\frac{1}{M} - \kappa\right)\left(\frac{1}{N} - \kappa\right) = 0 \implies \kappa_1 = \frac{1}{M}, \ \kappa_2 = \frac{1}{N}.$$

13.5   A curve on a surface

If the curve C runs on a curved surface, we may study some interesting things.

13.5.1   Tangent vector

If we call

$$t^i \stackrel{\text{def}}{=} \begin{bmatrix} t^1 \\ t^2 \end{bmatrix} = \begin{bmatrix} du/ds \\ dv/ds \end{bmatrix}$$

"the components of the vector $\mathbf{t}$ in the (u, v) co-ordinate system," this vector will be

$$\mathbf{t} = \frac{d\mathbf{x}}{ds} = \frac{d\mathbf{x}}{du}\frac{du}{ds} + \frac{d\mathbf{x}}{dv}\frac{dv}{ds} = t^1 \mathbf{x}_u + t^2 \mathbf{x}_v.$$

In other words, the tangent of the curve is also one tangent of the surface, and is located within the tangent plane which is spanned by $\mathbf{x}_u$ and $\mathbf{x}_v$.

13.5.2   Curvature vector

$$\mathbf{k} = \frac{d\mathbf{t}}{ds} = \frac{d}{ds}\left(t^1 \mathbf{x}_u + t^2 \mathbf{x}_v\right) =$$

$$= \mathbf{x}_u \frac{dt^1}{ds} + \mathbf{x}_v \frac{dt^2}{ds} + \mathbf{x}_{uu}\left(t^1\right)^2 + 2\mathbf{x}_{uv} t^1 t^2 + \mathbf{x}_{vv}\left(t^2\right)^2.$$

So, the "components of the curvature vector in the (u, v) co-ordinate system" contain something more than just the derivatives of the component values of the tangent vector, $\dfrac{dt^1}{ds}$ and $\dfrac{dt^2}{ds}$.

Let us write

$$
\begin{aligned}
\mathbf{x}_{uu} &= \Gamma_{11}^1 \mathbf{x}_u + \Gamma_{11}^2 \mathbf{x}_v + \langle \mathbf{x}_{uu} \cdot \mathbf{n} \rangle \, \mathbf{n}, \\
\mathbf{x}_{uv} &= \Gamma_{12}^1 \mathbf{x}_u + \Gamma_{12}^2 \mathbf{x}_v + \langle \mathbf{x}_{uv} \cdot \mathbf{n} \rangle \, \mathbf{n}, \\
\mathbf{x}_{vv} &= \Gamma_{22}^1 \mathbf{x}_u + \Gamma_{22}^2 \mathbf{x}_v + \langle \mathbf{x}_{vv} \cdot \mathbf{n} \rangle \, \mathbf{n};
\end{aligned}
\tag{13.8}
$$

i.e., expand the three-dimensional vectors on the frame $\{\mathbf{x}_u, \mathbf{x}_v, \mathbf{n}\}$ spanning the space.

Here appear — and show up in a natural way — the Γ symbols, also known as Christoffel symbols, which we shall discuss later (Chapter 14). They describe the reality that *differentiating a vector* (i.e., *parallel transport*, see Chapter 14) in curvilinear co-ordinates on a curved surface is *non-trivial*.

The third term in the equation's right-hand side however represents, according to Definition 13.4, *elements of the second fundamental form e, f, g*. We obtain

$$
\begin{aligned}
\mathbf{k} = {}& \left(\frac{dt^1}{ds} + \Gamma_{11}^1 \left(t^1\right)^2 + 2\Gamma_{12}^1 t^1 t^2 + \Gamma_{22}^1 \left(t^2\right)^2 \right) \mathbf{x}_u + \\
& + \left(\frac{dt^2}{ds} + \Gamma_{11}^2 \left(t^1\right)^2 + 2\Gamma_{12}^2 t^1 t^2 + \Gamma_{22}^2 \left(t^2\right)^2 \right) \mathbf{x}_v + \\
& + \left(e \left(t^1\right)^2 + 2f t^1 t^2 + g \left(t^2\right)^2 \right) \mathbf{n}.
\end{aligned}
$$

Here, the first two terms represent the *interior curvature* of curve C, $\mathbf{k}_{\text{int}}$, the curvature inside surface S; the last term represents the *exterior curvature* $\mathbf{k}_{\text{ext}}$, the bending of the curve with the curved surface itself.

The interior curvature again has two components in "(u, v) co-ordinates," which are obtained from the above equation. We write

$$
\mathbf{k}_{\text{int}} = k^1 \mathbf{x}_u + k^2 \mathbf{x}_v,
$$

in which

$$
k^i \stackrel{\text{def}}{=} \begin{bmatrix} k^1 \\ k^2 \end{bmatrix} =
\begin{bmatrix}
\dfrac{dt^1}{ds} + \Gamma_{11}^1 \left(t^1\right)^2 + 2\Gamma_{12}^1 t^1 t^2 + \Gamma_{22}^1 \left(t^2\right)^2 \\[2mm]
\dfrac{dt^2}{ds} + \Gamma_{11}^2 \left(t^1\right)^2 + 2\Gamma_{12}^2 t^1 t^2 + \Gamma_{22}^2 \left(t^2\right)^2
\end{bmatrix} =
$$

$$
=
\begin{bmatrix}
\dfrac{dt^1}{ds} + \displaystyle\sum_{i=1}^{2} \sum_{j=1}^{2} \Gamma_{ij}^1 t^i t^j \\[4mm]
\dfrac{dt^2}{ds} + \displaystyle\sum_{i=1}^{2} \sum_{j=1}^{2} \Gamma_{ij}^2 t^i t^j
\end{bmatrix},
$$

in which we have "economized" the formulas by using the summation sign. The exterior curvature is again

$$
\mathbf{k}_{\text{ext}} = \langle \mathbf{k} \cdot \mathbf{n} \rangle \, \mathbf{n},
$$

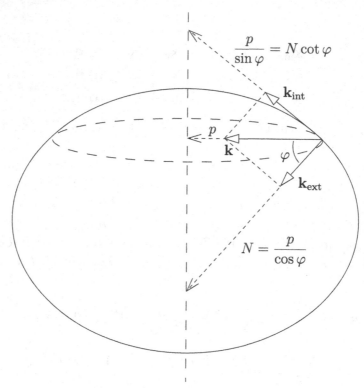

Figure 13.5: Curvatures and radii of curvature of latitude circles.

in which

$$\langle \mathbf{k} \cdot \mathbf{n} \rangle = e\left(t^1\right)^2 + 2f t^1 t^2 + g\left(t^2\right)^2.$$

Example. A latitude circle $\varphi = $ constant on the Earth's surface.

$$\mathbf{x} = \begin{bmatrix} N(\varphi)\cos\varphi\cos\lambda \\ N(\varphi)\cos\varphi\sin\lambda \\ N(\varphi)\left(1 - e^2\right)\sin\varphi \end{bmatrix};$$

$$\mathbf{t} = \frac{d\mathbf{x}}{ds} = \frac{d\mathbf{x}}{d\lambda}\frac{d\lambda}{ds} = \begin{bmatrix} -N\cos\varphi\sin\lambda \\ +N\cos\varphi\cos\lambda \\ 0 \end{bmatrix}\frac{1}{N\cos\varphi} = \begin{bmatrix} -\sin\lambda \\ +\cos\lambda \\ 0 \end{bmatrix};$$

$$\mathbf{k} = \frac{d\mathbf{t}}{ds} = \frac{d\mathbf{t}}{d\lambda}\frac{d\lambda}{ds} = \begin{bmatrix} -\cos\lambda \\ -\sin\lambda \\ 0 \end{bmatrix}\frac{1}{N\cos\varphi}.$$

The normal vector (Equation 13.7 on page 191):

$$\mathbf{n} = -\begin{bmatrix} \cos\varphi\cos\lambda \\ \cos\varphi\sin\lambda \\ \sin\varphi \end{bmatrix},$$

So, the exterior curvature is

$$\mathbf{k}_{\text{ext}} = \langle \mathbf{k} \cdot \mathbf{n} \rangle \, \mathbf{n} = \frac{1}{N} \left(\cos^2 \lambda + \sin^2 \lambda \right) \mathbf{n} = \frac{\mathbf{n}}{N}.$$

Because the vector $\mathbf{n}$ is a unit vector, we may infer that the curve's exterior curvature is precisely the inverse of the transversal radius of curvature N, and directed to the interior. It is the same as the curvature of the surface along the direction of the curve.

The interior curvature is

$$\mathbf{k}_{\text{int}} = \mathbf{k} - \mathbf{k}_{\text{ext}} = \frac{1}{N} \begin{bmatrix} -\dfrac{\cos \lambda}{\cos \varphi} + \cos \varphi \cos \lambda \\[2mm] -\dfrac{\sin \lambda}{\cos \varphi} + \cos \varphi \sin \lambda \\[2mm] + \sin \varphi \end{bmatrix} = \frac{\tan \varphi}{N} \begin{bmatrix} -\cos \lambda \sin \varphi \\[1mm] -\sin \lambda \sin \varphi \\[1mm] +\cos \varphi \end{bmatrix}.$$

We obtain for the length of this vector[13]

$$\|\mathbf{k}_{\text{int}}\| = \frac{\tan \varphi}{N}.$$

In Figure 13.5 we see the curvature vector $\mathbf{k}$, length

$$k \overset{\text{def}}{=} \|\mathbf{k}\| = \frac{1}{N \cos \varphi} = \frac{1}{p(\varphi)};$$

its interior part $\mathbf{k}_{\text{int}}$, length

$$\|\mathbf{k}_{\text{int}}\| = k \sin \varphi = \frac{1}{p(\varphi)} \sin \varphi = \frac{\tan \varphi}{N(\varphi)};$$

and its exterior part $\mathbf{k}_{\text{ext}}$, length

$$\|\mathbf{k}_{\text{ext}}\| = k \cos \varphi = \frac{1}{p(\varphi)} \cos \varphi = \frac{1}{N(\varphi)}.$$

From the picture, we also see how the distance from the Earth's axis of rotation is, in every direction, whether $\mathbf{k}, \mathbf{k}_{\text{int}}$ or $\mathbf{k}_{\text{ext}}$, the inverse of curvature. This is also intuitively clear from the rotational symmetry.

13.6 The geodesic

The *geodesic* is on a curved surface the *stationary curve* connecting two points. Normally[14] this means that it is the shortest distance between points; all other curves connecting the points are longer.

---

[13]Yes, it goes to infinity at the poles!

[14]But not always! For example, in relativity theory, the geodesic between two events in space-time is the world line connecting the events, which, measured in "proper time," is the *longest*. This also explains the twin paradox: the twin brother who stays on Earth and does not accelerate or brake, consumes more living time than his spacefaring brother.

13.6.1   Describing the geodesic interiorly, in surface co-ordinates

The equation for the geodesic is obtained by requiring $\mathbf{k}_{\text{int}} = k^i = 0$, i.e., the curve *has no interior curvature* (the exterior curvature cannot be eliminated when the curve is on a curved surface):

$$
\frac{dt^i}{ds} + \sum_{j=1}^{2} \sum_{k=1}^{2} \Gamma^i_{jk} t^j t^k = 0. \tag{13.9}
$$

Whe shall further develop this approach later in Section 14.6 on page 218.

13.6.2   Describing the geodesic exteriorly, using vectors in space

An alternative, three-dimensional ("exterior") form is obtained by requiring that the geodesic *is only exteriorly curved*, i.e., by writing

$$
\frac{d\mathbf{t}}{ds} = \mathbf{k}_{\text{ext}} = \left(e \left(t^1 \right)^2 + 2f t^1 t^2 + g \left(t^2 \right)^2 \right) \mathbf{n} = \left(\begin{bmatrix} t^1 & t^2 \end{bmatrix} \begin{bmatrix} e & f \\ f & g \end{bmatrix} \begin{bmatrix} t^1 \\ t^2 \end{bmatrix} \right) \mathbf{n}.
$$

Because

$$
\langle \mathbf{t} \cdot \mathbf{x}_u \rangle = t^1 \langle \mathbf{x}_u \cdot \mathbf{x}_u \rangle + t^2 \langle \mathbf{x}_u \cdot \mathbf{x}_v \rangle = Et^1 + Ft^2,
$$
$$
\langle \mathbf{t} \cdot \mathbf{x}_v \rangle = t^1 \langle \mathbf{x}_v \cdot \mathbf{x}_u \rangle + t^2 \langle \mathbf{x}_v \cdot \mathbf{x}_v \rangle = Ft^1 + Gt^2,
$$

it follows that[15]

$$
\begin{bmatrix} t^1 \\ t^2 \end{bmatrix} = \begin{bmatrix} E & F \\ F & G \end{bmatrix}^{-1} \begin{bmatrix} \langle \mathbf{t} \cdot \mathbf{x}_u \rangle \\ \langle \mathbf{t} \cdot \mathbf{x}_v \rangle \end{bmatrix}.
$$

We may write

$$
\frac{d\mathbf{t}}{ds} = \left(\begin{bmatrix} \langle \mathbf{t} \cdot \mathbf{x}_u \rangle & \langle \mathbf{t} \cdot \mathbf{x}_v \rangle \end{bmatrix} \begin{bmatrix} E & F \\ F & G \end{bmatrix}^{-1} \cdot \begin{bmatrix} e & f \\ f & g \end{bmatrix} \begin{bmatrix} E & F \\ F & G \end{bmatrix}^{-1} \begin{bmatrix} \langle \mathbf{t} \cdot \mathbf{x}_u \rangle \\ \langle \mathbf{t} \cdot \mathbf{x}_v \rangle \end{bmatrix} \right) \mathbf{n}.
$$

Here

$$
\begin{bmatrix} E & F \\ F & G \end{bmatrix}^{-1} \begin{bmatrix} e & f \\ f & g \end{bmatrix} \begin{bmatrix} E & F \\ F & G \end{bmatrix}^{-1} = \mathbf{H}^{-1} \mathbf{B} \mathbf{H}^{-1}
$$

following the earlier used notation[16].

---

[15]This suggests the notation $\begin{bmatrix} t_1 \\ t_2 \end{bmatrix} \overset{\text{def}}{=} \begin{bmatrix} \langle \mathbf{t} \cdot \mathbf{x}_u \rangle \\ \langle \mathbf{t} \cdot \mathbf{x}_v \rangle \end{bmatrix}.$

[16]In index notation: $g^{ij} \beta_{jk} g^{k\ell}$, see Chapter 14. A logical notation for this object would be $\beta^{i\ell}$.

Example. On the surface of an ellipsoidal Earth, we have

$$\mathbf{H}^{-1}\mathbf{B}\mathbf{H}^{-1} = \begin{bmatrix} M^{-2} & 0 \\ 0 & N^{-2}\cos^{-2}\varphi \end{bmatrix} \begin{bmatrix} M & 0 \\ 0 & N\cos^2\varphi \end{bmatrix} \begin{bmatrix} M^{-2} & 0 \\ 0 & N^{-2}\cos^{-2}\varphi \end{bmatrix} =$$

$$= \begin{bmatrix} M^{-3} & 0 \\ 0 & N^{-3}\cos^{-2}\varphi \end{bmatrix} = \begin{bmatrix} \dfrac{1}{M^3} & 0 \\ 0 & \dfrac{\cos\varphi}{p^3} \end{bmatrix}$$

and

$$\begin{bmatrix} \langle \mathbf{t}\cdot\mathbf{x}_u \rangle \\ \langle \mathbf{t}\cdot\mathbf{x}_v \rangle \end{bmatrix} = \begin{bmatrix} M\langle \mathbf{t}\cdot\mathbf{e}_N \rangle \\ p\langle \mathbf{t}\cdot\mathbf{e}_E \rangle \end{bmatrix},$$

from which we obtain

$$\frac{d\mathbf{t}}{ds} = \begin{bmatrix} \langle \mathbf{t}\cdot\mathbf{e}_N \rangle & \langle \mathbf{t}\cdot\mathbf{e}_E \rangle \end{bmatrix} \begin{bmatrix} \dfrac{1}{M} & 0 \\ 0 & \dfrac{\cos\varphi}{p} \end{bmatrix} \begin{bmatrix} \langle \mathbf{t}\cdot\mathbf{e}_N \rangle \\ \langle \mathbf{t}\cdot\mathbf{e}_E \rangle \end{bmatrix} \mathbf{n} =$$

$$= \left(\frac{1}{M}\langle \mathbf{t}\cdot\mathbf{e}_N \rangle^2 + \frac{1}{N}\langle \mathbf{t}\cdot\mathbf{e}_E \rangle^2 \right)\mathbf{n}. \tag{13.10}$$

Here, the unit vectors in the North and East directions are

$$\mathbf{e}_N = \frac{\mathbf{x}_u}{\|\mathbf{x}_u\|} = \begin{bmatrix} -\sin\varphi\cos\lambda \\ -\sin\varphi\sin\lambda \\ \cos\varphi \end{bmatrix}, \quad \mathbf{e}_E = \frac{\mathbf{x}_v}{\|\mathbf{x}_v\|} = \begin{bmatrix} -\cos\varphi\sin\lambda \\ +\cos\varphi\cos\lambda \\ 0 \end{bmatrix}.$$

The approach is geometrically intuitive. The expression in the curly brackets is, with A the azimuth,

$$\frac{1}{M}\langle \mathbf{t}\cdot\mathbf{e}_N \rangle^2 + \frac{1}{N}\langle \mathbf{t}\cdot\mathbf{e}_E \rangle^2 = \frac{1}{M}\cos^2 A + \frac{1}{N}\sin^2 A$$

precisely *the curvature of the surface in the direction of the geodesic.*

In addition to Equations 13.10, we have still the equations defining the tangent vector

$$\frac{d\mathbf{x}}{ds} = \mathbf{t},$$

altogether $2 \times 3 = 6$ ordinary differential equations.

The space method has both *advantages* and *disadvantages* compared to the method of surface co-ordinates (e.g., Equations 8.1 on page 92).

Advantages: In surface co-ordinates (φ, λ) there are inevitably always two *poles*, singularities, where the curvature of the latitude circles goes to infinity, and numerical methods may behave badly. This does not happen in rectangular space co-ordinates.

Disadvantages:

1. More equations means more computational work.

2. In every point one has to compute M and N, and for these, $\varphi = \arctan \dfrac{Z}{(1 - e^2)\, p}$, in which $p = \sqrt{X^2 + Y^2}$.

3. The roll-in and roll-out of the computation requires the transformation of (φ, λ) to (X, Y, Z) co-ordinates in the starting point, and back again in the end point.

13.7   More generally on surface theory

In the foregoing chapter, as well as in the following chapter on Riemann's surface theory, we study surface theory from the *local*, i.e., the differential-geometry, *viewpoint*. Here we study only small areas. If we look at large surfaces in $\mathbb{R}^3$ in their totality, we meet complications. For example, already parametrizing the globe using latitude and longitude (φ, λ) gives rise to two "poles," singular points, that one cannot get rid of in any way by changing parametrization. For this reason, the polar areas are often mapped separately in their own projection.

Mathematically, a surface can be imagined as the graph of a scalar field of function $\phi \colon D \to \mathbb{R}$, when D is a domain in the plane $\mathbb{R}^2$. The problem however is, that, in the cases that are interesting from the viewpoint of practical applications — e.g., on the surface of a sphere — the function ϕ cannot be chosen such, that it would be unique. Therefore, the surface must generally be presented in pieces; e.g., in the case of a sphere, the top and bottom halves of the sphere must be presented separately. Then, one must take care that the separate pieces presenting the surface all fit together.

The idea is the same as in an atlas, i.e., a map set consisting of many sheets, where the separate map sheets together form a presentation of a global map. Like in the case of an atlas, also in the mathematical version of the idea, one places the adjoining map sheets slightly overlappingly so that when changing the page one knows how the map ought to continue on the next page. The function for an individual map sheet is also called, in the mathematical version, the chart ϕ_j, and the surface presentation formed by the charts together is again called an "atlas."

When choosing charts, one should however still take care, that the presentation that results is mathematically sufficiently well behaved with regard to its conformity and other properties useful in studying map projections. Therefore, it is assumed that the ϕ_j charts are injections (i.e., the inverse chart

ϕ_j^{-1} is defined for all j) and functions of the form $\phi_j \circ \phi_k^{-1}$ are analytic on their domains.

Formally, a Riemann surface is made up of the following elements:

- a subset of Euclidean space[17], X

- a collection of open sets $U_j \subset X$, that together form a cover of set X (i.e., $X = \cup_j U_j$)

- for all values j, the chart

$$\phi_j \colon U_j \to \phi(U_j)$$

is a homeomorphism[18] and the composition chart $\phi_j \circ \phi_k^{-1}$ is analytic for all values k in all the points where it is defined.

This definition will probably seem hard to understand. In practice, however, it only means that the surface is presented in neat pieces that fit together. In practice, the choice of the mappings is not generally difficult.

The significance of Riemann surfaces lies in the ability to mathematically define and study, e.g., conformal mappings on other two-dimensional surfaces than the complex plane. When using charts, their definition reverts to the study of composition charts. Charts are a mathematical device, the choice of which does not affect the end result.

Exercises

Exercise 13 – 1: The metric of spacetime

In relativity theory, a point in spacetime is called an *event*

$$\mathbf{x} = \begin{bmatrix} x & y & z & t \end{bmatrix}^\mathsf{T}.$$

The square of the distance or *interval* between two events is:

$$\|\mathbf{x}_2 - \mathbf{x}_1\|^2 = c^2(t_2 - t_1)^2 - (x_2 - x_1)^2 - (y_2 - y_1)^2 - (z_2 - z_1)^2.$$

This is an *invariant*, i.e., it is the same independent of the choice of co-ordinate system (which we here assume however to be rectangular and moving uniformly relative to each other; i.e., "base co-ordinates" in spacetime). Let us write it in differential form:

$$ds^2 = c^2 dt^2 - dx^2 - dy^2 - dz^2,$$

which defines the *metric* for the four-dimensional spacetime.

---

[17]More generally, a so-called Hausdorff topological space.

[18]ϕ_j is continuous and its inverse chart ϕ_j^{-1} is defined and continuous.

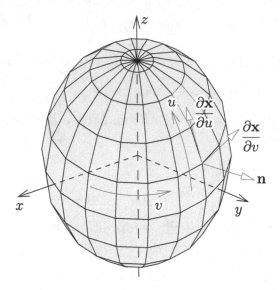

Figure 13.6: Triaxial ellipsoid, tangent vectors and (exterior) normal.

a. Write out the *metric tensor* $g_{ij}, i, j = 1, \ldots, 4$ (Minkowski metric)

b. Consider *only* the (x, t) plane. Write up the Gauss fundamental quantities E, F, G.

c. Can ds^2 be negative?

d. Does the inference $\|\mathbf{x}_2 - \mathbf{x}_1\| = 0 \implies \mathbf{x}_1 = \mathbf{x}_2$ hold?

Exercise 13 – 2: The metric of a triaxial ellipsoid

Given a *triaxial* ellipsoid (Figure 13.6):

$$\frac{x^2}{a^2} + \frac{y^2}{b^2} + \frac{z^2}{c^2} = 1.$$

On the surface of the ellipsoid, we create a curvilinear co-ordinate system:

$$x = a \cos u \cos v$$
$$y = b \cos u \sin v$$
$$z = c \sin u$$

a. Calculate the vectors

$$\mathbf{x}_u \stackrel{\text{def}}{=} \frac{\partial \mathbf{x}}{\partial u} = \begin{bmatrix} \partial x / \partial u \\ \partial y / \partial u \\ \partial z / \partial u \end{bmatrix}, \ \mathbf{x}_v = \begin{bmatrix} \partial x / \partial v \\ \partial y / \partial v \\ \partial z / \partial v \end{bmatrix}$$

b. Calculate the Gauss first fundamental quantities E, F, G, and form the
   metric tensor

$$g_{ij} = \mathbf{H} = \begin{bmatrix} E & F \\ F & G \end{bmatrix}.$$

Chapter 14

Riemann surfaces and charts

An important theoretical frame in which curved surfaces and curves are often decribed is *Riemann's surface theory.* Georg Friedrich Bernhard Riemann (1826 – 1866), a brilliant German mathematician in his own right, was a student of Gauss. See Figure 14.1. He is remembered not only for his contributions to differential geometry, but also for his achievements in number theory.

Unlike Gauss, Riemann was otherwise a balanced person, but he was shy and suffered from health problems. Like Gauss, also Riemann was reluctant to publish work that he considered unfinished. Riemann spent the final part of his life in Italy, where he died of tuberculosis at only 40 years of age. He is buried in Biganzolo, Verbania, by the Lago Maggiore.

In Riemann's theoretical frame, one studies a curved surface *intrinsically*, i.e., without considering that the curved surface of the Earth is "embedded" in a whole three-dimensional (Euclidean) space.

This makes Riemann's surface theory a useful abstraction also in situations where the surrounding higher dimensional space does not necessarily even exist. For example, in the general theory of relativity, we describe spacetime (x, y, z, t) as a curved continuum conforming to Riemann's theory, the curvature parameters of which are associated with the densities of masses and flows of matter as described in Einstein[1]'s field equations. The gravitational field is the expression of this curvature.

Figure 14.1: Georg Friedrich Bernhard Riemann, taken in 1863 (Wikimedia Commons, Riemann, portrait).

A deeper study of Riemann's surface theory requires a bit of vector and tensor calculus as well as an understanding of tensor notation. Tensor notation and its interpretation is closely linked to the concept of *invariance*. In physics, the invariance of a quantity means that

---

[1] Albert Einstein (1879 – 1955) was a German-born theoretical physicist, discoverer of relativity theory, reluctant quantum theoretician and Nobel laureate, peace activist and icon of science.

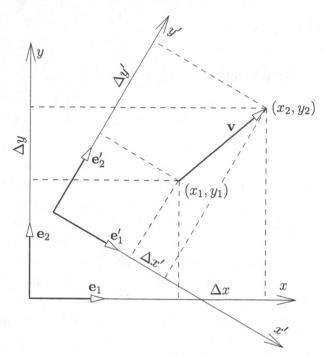

Figure 14.2: The vector $\mathbf{v} = \mathbf{x}_2 - \mathbf{x}_1$ in the plane described on two different co-ordinate frames. The vector pairs $\{\mathbf{e}_1, \mathbf{e}_2\}, \{\mathbf{e}'_1, \mathbf{e}'_2\}$ are the orthonormal bases for the frames.

the form of the formula for its computation does not change, even though the computations were performed in different co-ordinate frames.

A familiar example of invariance is the distance between two points: e.g., the length of a material object, like a ruler. When the object is arbitrarily moved around, its measures do not change, which is the physical justification behind this invariance concept. Let there be two points $\mathbf{x}_1$, co-ordinates x_1, y_1, z_1 and $\mathbf{x}_2$, co-ordinates x_2, y_2, z_2 on the object. Then, in a rectangular (Cartesian) co-ordinate frame, the distance between the points is

$$\|\mathbf{x}_2 - \mathbf{x}_1\| = \sqrt{(x_2 - x_1)^2 + (y_2 - y_1)^2 + (z_2 - z_1)^2}.$$

In another rectangular co-ordinate frame, *the same distance* is

$$\|\mathbf{x}_2 - \mathbf{x}_1\| = \sqrt{(x'_2 - x'_1)^2 + (y'_2 - y'_1)^2 + (z'_2 - z'_1)^2}.$$

The expressions are identical in form, although the co-ordinates themselves are different.

In physics, there are many invariants. For example, the speed of light c is an invariant, when the set of co-ordinate frames under consideration consists of so-called *Lorentz frames*, the extension of the set of rectangular co-ordinate frames in space to the proper co-ordinate frames of observers in spacetime moving at constant speeds.

In Riemann's surface theory, the set of co-ordinate systems under consideration is still broader: it comprises all curvilinear (but "well behaved") co-ordinate frames on a curved surface. This is the starting point of the theory.

A nice, slow introduction to Riemann surface theory and its application to relativity theory is found in Carroll (1997). Another nice introduction, though more theoretical, is Lee (1997).

14.1 What is a tensor?

In physics we encounter, more often than in geodesy, the concept of the *tensor*. What is a tensor?

14.1.1 Vectors

Let us first look at the situation in two-dimensional Euclidean space, i.e., the *plane* $\mathbb{E}^2$. See Figure 14.2. The pair of co-ordinate differences between two adjacent points corresponds to the vector

$$\mathbf{v} = \sum_{i=1}^{2} v^i \mathbf{e}_i, \quad \text{where} \quad v^i = \begin{bmatrix} \Delta x \\ \Delta y \end{bmatrix} = \begin{bmatrix} x_2 - x_1 \\ y_2 - y_1 \end{bmatrix}$$

is the vector expressed using superscripts, and $\mathbf{e}_i, i = 1 \ldots 2$ is a suitable orthonormal basis of $\mathbb{E}^2$.

In another co-ordinate frame, this same vector is written

$$\mathbf{v} = \sum_{i=1}^{2} v^{i'} \mathbf{e}_{i'}, \quad \text{where} \quad v^{i'} = \begin{bmatrix} \Delta x' \\ \Delta y' \end{bmatrix} = \begin{bmatrix} x_2' - x_1' \\ y_2' - y_1' \end{bmatrix}$$

and $\mathbf{e}_{i'}$ is another orthonormal basis of $\mathbb{E}^2$.

For the vectors we use both a symbolic ($\mathbf{v}$) and a component-wise (v^i, $i = 1, 2$) notation. The components will depend on the chosen co-ordinate frame ($\Delta x' \neq \Delta x$, $\Delta y' \neq \Delta y$, though always $\Delta x^2 + \Delta y^2 = (\Delta x')^2 + (\Delta y')^2 = \text{constant}$, i.e., an *invariant*), but the symbolic notation does not depend on that. The vector is always the same, even though its components transform with the co-ordinate frame used. The vector is always the same "arrow in space."

For transforming the components of the vector, the following transformation formula exists:

$$v^{i'} = \sum_i \alpha_i^{i'} v^i, \tag{14.1}$$

or, as a matric equation,

$$\overline{v}' = A\overline{v},$$

in which

$$\overline{v} = v^i = \begin{bmatrix} v^1 \\ v^2 \end{bmatrix} \left(= \begin{bmatrix} \Delta x \\ \Delta y \end{bmatrix} \right)$$

is now a *column matrix of components*[2], i.e., an element of $\mathbb{R}^2$, and

$$A = \alpha_i^{i'} = \begin{array}{c} i' \downarrow \\ \begin{bmatrix} \alpha_1^1 & \alpha_2^1 \\ \alpha_1^2 & \alpha_2^2 \end{bmatrix} \\ i \rightarrow \end{array} \left(= \begin{bmatrix} \cos\theta & \sin\theta \\ -\sin\theta & \cos\theta \end{bmatrix} \right)$$

is the component matrix of the transformation operator. As seen, i is a column index and i' a row index. θ stands for the rotation from the one co-ordinate frame to the other, reckoned positive counter-clockwise.

Every quantity that transforms according to Equation 14.1, we call a *vector*. Familiar vectorial quantities are, e.g., speed, acceleration and force. What they have in common is, that they may be graphically presented as *arrows*.

14.1.2 Tensors

After this, a *tensor* is defined as a kind of matrix, which behaves in the same way under transformation from one co-ordinate frame to another, but *separately for each index*:

$$T^{i'j'} = \sum_{i,j} \alpha_i^{i'} \alpha_j^{j'} T^{ij}.$$

As a matric equation, this is[3]

$$\overline{T}' = A\overline{T}A^{\mathsf{T}},$$

in which A is the same as above.

In geodesy, we use many tensors:

1. The Earth's inertial tensor.

2. The gravity-gradient tensor

$$M = \begin{bmatrix} \partial g/\partial x \\ \partial g/\partial y \\ \partial g/\partial z \end{bmatrix} = \begin{bmatrix} \dfrac{\partial^2 W}{\partial x^2} & \dfrac{\partial^2 W}{\partial x \partial y} & \dfrac{\partial^2 W}{\partial x \partial y} \\ \dfrac{\partial^2 W}{\partial y \partial x} & \dfrac{\partial^2 W}{\partial y^2} & \dfrac{\partial^2 W}{\partial x \partial z} \\ \dfrac{\partial^2 W}{\partial z \partial x} & \dfrac{\partial^2 W}{\partial z \partial y} & \dfrac{\partial^2 W}{\partial z^2} \end{bmatrix}.$$

---

[2]In the sequel we will leave the overbar away when no risk of confusion exists.

[3]Verify that the matric equation leads to the same index summations as the index equation!

3. The variance matrix is really a tensor: $\text{Var}\{\mathbf{x}\} = \begin{bmatrix} \sigma_x^2 & \sigma_{xy} \\ \sigma_{xy} & \sigma_y^2 \end{bmatrix}$, where σ_x and σ_y are the mean errors of the co-ordinates, i.e., of the components of $\mathbf{x} = \begin{bmatrix} x & y \end{bmatrix}^{\mathsf{T}}$, and σ_{xy} is the covariance between them.

4. The first, $\mathbf{H} = \begin{bmatrix} E & F \\ F & G \end{bmatrix}$, and the second, $\mathbf{B} = \begin{bmatrix} e & f \\ f & g \end{bmatrix}$, fundamental form of the surface theory of Gauss, as well as $\mathbf{C} = \mathbf{H}^{-1}\mathbf{B}$, all discussed in Section 13.4.

5. Also the elements of the fundamental form of map projection (on which more in the next chapter) $\widetilde{E}, \widetilde{F}, \widetilde{G}$ form a tensor $\widetilde{\mathbf{H}} = \begin{bmatrix} \widetilde{E} & \widetilde{F} \\ \widetilde{F} & \widetilde{G} \end{bmatrix}$, which is a kind of metric tensor.

The object $\mathbf{H}^{-1}\widetilde{\mathbf{H}} = \begin{bmatrix} \widetilde{E}/M^2 & \widetilde{F}/Mp \\ \widetilde{F}/Mp & \widetilde{G}/p^2 \end{bmatrix}$ again could be called the *scale tensor*.

14.1.3 The geometric presentation of a tensor and its invariants

In the same way that the geometric depiction of a *vector* is an *arrow*, also the geometric depiction of a *tensor* is an *ellipse* (two-dimensionally) or an *ellipsoid* (three-dimensionally)[4]. The lengths of the principal axes of the ellipse or ellipsoid describe the eigenvalues[5] of the tensor. The directions of the principal axes again are the directions of the tensor's eigenvectors.

In Euclidean space and in basic rectangular co-ordinates, square tensors T^{ij} are generally *symmetric*. Therefore, the eigenvectors $\mathbf{x}_i, \mathbf{x}_j$ belonging to different eigenvalues $\lambda_i, \lambda_j, i \neq j$ are mutually orthogonal, as proven in mathematics textbooks. See Subsection 14.3.2 for a deeper discussion.

One can show that in an n-dimensional space, a tensor T has n *independent invariants*. The eigenvalues λ_i, $i = 1, \ldots, n$ are of course invariants. So are also their *sum* and *product*,

$$\sum_i \lambda_i = \sum_i T^{ii},$$

in two dimensions

$$\lambda_1 + \lambda_2 = T^{11} + T^{22},$$

the sum of the diagonal elements or *trace*[6], and

$$\prod_i \lambda_i = \det(\mathbf{T}),$$

---

[4] Actually only if the tensor is positive definite, i.e., all its different eigenvalues are positive.
[5] More precisely, the lengths of the half-axes are the square roots of the eigenvalues λ_i.
[6] German: *Spur*.

two-dimensionally again

$$\lambda_1\lambda_2 = \det\left(\mathbf{T}\right) = T^{11}T^{22} - T^{12}T^{21},$$

the *determinant* of the tensor.

The trace of a variance matrix $\sigma_x^2 + \sigma_y^2$ is known in geodesy as the *point variance* σ_P^2. It is chosen precisely because it is an invariant, i.e., independent of the directions of the x and y axes. Also the trace of the gravity-gradient tensor has meaning:

$$\sum_i M_{ii} = \frac{\partial^2 W}{\partial x^2} + \frac{\partial^2 W}{\partial y^2} + \frac{\partial^2 W}{\partial z^2} \overset{\text{def}}{=} \Delta W,$$

the *Laplace operator* acting on the geopotential.

14.1.4   Tensors in general co-ordinates

What happens in a *non-Euclidean co-ordinate frame*? In this case the distinction between super- and subscripts turns out to be meaningful and finally finds its justification.

A *contravariant vector* transforms as follows:

$$v^{i'} = \sum_i \alpha_i^{i'} v^i \tag{14.2}$$

and a *covariant vector* as follows:

$$v_{i'} = \sum_i \alpha_{i'}^{i} v_i. \tag{14.3}$$

(Very similar looking, but not the same!)

Whereas the "prototype" of a contravariant vector is given by the co-ordinate differences between two points (close to each other), like above,

$$v^i = \left[\begin{array}{c} \Delta x \\ \Delta y \end{array}\right],$$

the prototype of a covariant vector is the *gradient operator*:

$$v_j = \frac{\partial V}{\partial x^j} = \left[\begin{array}{c} \partial V/\partial x \\ \partial V/\partial y \end{array}\right], \tag{14.4}$$

where $V\left(x, y\right)$ is some scalar field in space.

If we take as a model vector

$$v^i = \left[\begin{array}{c} dx \\ dy \end{array}\right],$$

we obtain

$$v^{i'} = \begin{bmatrix} dx' \\ dy' \end{bmatrix} = \begin{bmatrix} \partial x'/\partial x & \partial x'/\partial y \\ \partial y'/\partial x & \partial y'/\partial y \end{bmatrix} \begin{bmatrix} dx \\ dy \end{bmatrix} ;$$

also, with the chain rule, with v_j as in Equation 14.4:

$$v_{j'} = \begin{bmatrix} \partial V/\partial x' \\ \partial V/\partial y' \end{bmatrix} = \begin{bmatrix} \partial x/\partial x' & \partial y/\partial x' \\ \partial x/\partial y' & \partial y/\partial y' \end{bmatrix} \begin{bmatrix} \partial V/\partial x \\ \partial V/\partial y \end{bmatrix} .$$

The coefficient matrices $\dfrac{\partial x^{i'}}{\partial x^i} = \alpha_i^{i'}$ and $\dfrac{\partial x^i}{\partial x^{i'}} = \alpha_{i'}^i$ are, by inspection, *each other's transposed inverse matrices*.

Thus, if the matrix notation for the transformation parameters $\alpha_i^{i'}$ of the contravariant transformation formula 14.2 is $\mathbf{A}$ (column index i and row index i'), the matrix for the covariant transformation parameters (formula 14.3) $\alpha_{i'}^i$ is $\left(\mathbf{A}^{\mathsf{T}}\right)^{-1}$ (column index i and row index i'). This is the origin of the namings covariant and contravariant[7].

Tensors can have both upper and lower indices, i.e., super- and subscripts. When changing the co-ordinate frame, each index transforms according to its "nature." The tensor does not even necessarily have to have only two indices — there may be more of them.

14.1.5   Trivial tensors

1. If we compute the gradient of the contravariant vector x^i, we obtain

$$\frac{\partial x^i}{\partial x^j} = \begin{bmatrix} 1 & 0 \\ 0 & 1 \end{bmatrix} \stackrel{\text{def}}{=} \delta_j^i ,$$

the so-called Kronecker[8] delta, which in practice is the unit or identity matrix. It is a tensor:

$$\delta_{j'}^{i'} = \alpha_{j'}^j \alpha_i^{i'} \delta_j^i = \sum_{i,j} \alpha_i^{i'} \delta_j^i \alpha_{j'}^j ,$$

or, in matrix language,

$$\begin{array}{cccc} i'\downarrow & i'\downarrow & i\downarrow & j\downarrow \\ [\mathbf{I}'] \;=\; & [\mathbf{A}] & [\mathbf{I}] & [\mathbf{A}^{-1}] = \mathbf{A}\mathbf{A}^{-1} = \mathbf{I}, \\ j'\rightarrow & i\rightarrow & j\rightarrow & j'\rightarrow \end{array}$$

---

[7]If the co-ordinate frames are rectangular (Cartesian), all transformation matrices are orthogonal, i.e., their transposed inverse equals the original matrix, see Section 9.3. That is why, in this situation, the distinction between co- and contravariant is meaningless.

[8]Leopold Kronecker (1823 – 1891) was a German mathematician who advanced number theory and algebra.

because, if the matrix presentation of $\alpha_i^{i'}$ is $\mathbf{A}$, then the matrix presentation of $\alpha_{j'}^{j}$ — correspondingly that of $\alpha_{i'}^{i}$ — is $\left(\mathbf{A}^\mathsf{T}\right)^{-1}$, as we concluded above.

2. The Levi-Civita[9] "corkscrew tensor" in three dimensions:

$$\epsilon_{ijk} = \begin{cases} 0 & \text{if two of } ijk \text{ are the same} \\ 1 & \text{if } ijk \text{ an even permutation of the numbers (123)} \\ -1 & \text{if } ijk \text{ an odd permutation of the numbers (123)}. \end{cases}$$

In other words, $\epsilon_{123} = \epsilon_{231} = \epsilon_{312} = 1$, $\epsilon_{132} = \epsilon_{321} = \epsilon_{213} = -1$, all others $= 0$.

See Wikipedia, Levi-Civita symbol.

Kronecker and Levi-Civita are also called *isotropic* tensors, because they do not change when rotating the co-ordinate axes.

14.2 The metric tensor

The metric tensor, or *metric*, describes the form of the Pythagoras theorem in curved space. It is the same as the earlier discussed Gauss first fundamental form. In the ordinary plane ($\mathbb{R}^2$) we may choose rectangular co-ordinates (x, y), after which the distance s between two points $1, 2$ can be written as

$$s^2 = \Delta x^2 + \Delta y^2,$$

where $\Delta x = x_2 - x_1, \Delta y = y_2 - y_1$ are the co-ordinate differences between the points. This formula applies in the whole plane. Its differential version looks similar:

$$ds^2 = dx^2 + dy^2.$$

This is now written in the following form:

$$ds^2 = g_{ij} \, dx^i dx^j, \tag{14.5}$$

where

$$g_{ij} = \begin{bmatrix} 1 & 0 \\ 0 & 1 \end{bmatrix} \quad \text{and} \quad dx^i = dx^j = \begin{bmatrix} dx^1 \\ dx^2 \end{bmatrix} = \begin{bmatrix} dx \\ dy \end{bmatrix}.$$

---

[9]Tullio Levi-Civita (1873–1941) was an Italian mathematician, who, together with his teacher Gregorio Ricci-Curbastro, further developed tensor calculus and published an important textbook, which also Einstein used. Einstein is quoted as saying that two good things came from Italy: spaghetti and Levi-Civita.

This is referred to as the *metric* of rectangular or Cartesian co-ordinates in the Euclidean plane. In Equation 14.5 it has been assumed that we sum over the indices i and j. This assumption is called the *Einstein summation convention*. Always when a formula contains an identical super- and subscript, we sum over it. So, in this case

$$ds^2 = \sum_{i=1}^{2} \sum_{j=1}^{2} g_{ij} \, dx^i dx^j.$$

An alternative way to write this using matrix notation is the square of a *norm*, or the *quadratic form*

$$ds^2 = \|\mathbf{x}\|_{\mathbf{H}}^2 = \langle \mathbf{x} \cdot \mathbf{x} \rangle_{\mathbf{H}} = \mathbf{x}^{\mathsf{T}} \mathbf{H} \mathbf{x} =$$

$$= \overbrace{\begin{bmatrix} dx^1 & dx^2 \end{bmatrix}}^{\mathbf{x}^{\mathsf{T}}} \overbrace{\begin{bmatrix} g_{11} & g_{12} \\ g_{21} & g_{22} \end{bmatrix}}^{\mathbf{H}} \overbrace{\begin{bmatrix} dx^1 \\ dx^2 \end{bmatrix}}^{\mathbf{x}}.$$

If the surface is not curved, we can always find a co-ordinate frame that is *everywhere* rectangular and both co-ordinates scaled such that to a co-ordinate difference of 1 m corresponds also a difference in place of 1 m. Then, the g_{ij} matrix or *metric tensor* is of the form of the unit matrix, like above.

If the surface is curved, we may find a unit matrix *only* in some points of the plane. For example, on the surface of the unit sphere parametrized by spherical co-ordinates (ϕ, λ) this is possible only on the equator of the parametrization (latitude $\phi = 0$). It is not possible contiguously and precisely on the whole sphere, and not even on any part of its surface.

We may however choose, even on a non-curved plane, a co-ordinate frame that is not straight but *skewed*, and where the co-ordinates are arbitrarily scaled. See Figure 14.3. In this case, the cosine rule in triangle $\triangle ABC$ gives us

$$s^2 = p^2 \Delta u^2 + q^2 \Delta v^2 + 2p \, \Delta u \, q \, \Delta v \cos \alpha,$$

or, differentially,

$$ds^2 = p^2 du^2 + q^2 dv^2 + 2pq \cos \alpha \, du dv,$$

or, in index notation,

$$ds^2 = g_{ij} \, dx^i dx^j$$

with

$$g_{ij} = \begin{bmatrix} p^2 & pq \cos \alpha \\ pq \cos \alpha & q^2 \end{bmatrix}, \, dx^i = \begin{bmatrix} du \\ dv \end{bmatrix}.$$

In matrix presentation:

$$ds^2 = \overbrace{\begin{bmatrix} du & dv \end{bmatrix}}^{dx^i} \overbrace{\begin{bmatrix} p^2 & pq \cos \alpha \\ pq \cos \alpha & q^2 \end{bmatrix}}^{g_{ij}, \downarrow i, \rightarrow j} \overbrace{\begin{bmatrix} du \\ dv \end{bmatrix}}^{dx^j}.$$

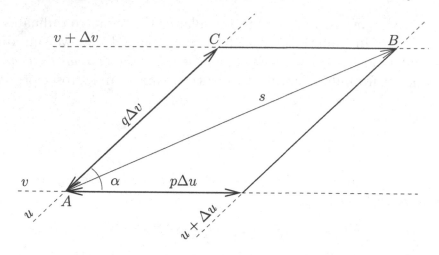

Figure 14.3: A skewed metric in the plane.

On a *curved* surface, g_{ij} depends on place: $g_{ij}(x^i)$, where x^i is the formal vector of parameters describing the surface, $x^i = \begin{bmatrix} u & v \end{bmatrix}^\mathsf{T}$. Also on a non-curved surface, g_{ij} may be dependent on place, e.g., if we choose curvilinear co-ordinates, e.g., polar co-ordinates.

Example (1). The surface of a spherical Earth:

$$ds^2 = R^2 d\varphi^2 + R^2 \cos^2 \varphi \, d\lambda^2,$$

or

$$g_{ij}(\varphi, \lambda) = \begin{bmatrix} R^2 & 0 \\ 0 & R^2 \cos^2 \varphi \end{bmatrix}, \quad dx^i = \begin{bmatrix} d\varphi \\ d\lambda \end{bmatrix},$$

so that, as a matric equation

$$ds^2 = \overbrace{\begin{bmatrix} d\varphi & d\lambda \end{bmatrix}}^{dx^i} \overbrace{\begin{bmatrix} R^2 & 0 \\ 0 & R^2 \cos^2 \varphi \end{bmatrix}}^{g_{ij},\, \downarrow i,\, \rightarrow j} \overbrace{\begin{bmatrix} d\varphi \\ d\lambda \end{bmatrix}}^{dx^j}$$

Example (2). Polar co-ordinates in the plane (ρ, θ):

$$ds^2 = d\rho^2 + \rho^2 d\theta^2,$$

or

$$g_{ij} = \begin{bmatrix} 1 & 0 \\ 0 & \rho^2 \end{bmatrix}, \quad dx^i = \begin{bmatrix} d\rho \\ d\theta \end{bmatrix},$$

giving

$$ds^2 = \overbrace{\begin{bmatrix} d\rho & d\theta \end{bmatrix}}^{dx^i} \overbrace{\begin{bmatrix} 1 & 0 \\ 0 & \rho^2 \end{bmatrix}}^{g_{ij},\, \downarrow i,\, \rightarrow j} \overbrace{\begin{bmatrix} d\rho \\ d\theta \end{bmatrix}}^{dx^j}.$$

Example (3). *Three-dimensionally* in the air space, aviation co-ordinates are: nautical miles North dN, nautical miles East dE, feet above sea level dH; in metres

$$ds^2 = (1852)^2 \, dN^2 + (1852)^2 \, dE^2 + (0.3048)^2 \, dH^2,$$

or

$$g_{ij} = \begin{bmatrix} 1852^2 & 0 & 0 \\ 0 & 1852^2 & 0 \\ 0 & 0 & 0.3048^2 \end{bmatrix}, dx^i = \begin{bmatrix} dN \\ dE \\ dH \end{bmatrix},$$

and

$$ds^2 = \overbrace{\begin{bmatrix} dN & dE & dH \end{bmatrix}}^{dx^i} \overbrace{\begin{bmatrix} 1852^2 & 0 & 0 \\ 0 & 1852^2 & 0 \\ 0 & 0 & 0,3048^2 \end{bmatrix}}^{g_{ij}, \downarrow i, \rightarrow j} \overbrace{\begin{bmatrix} dN \\ dE \\ dH \end{bmatrix}}^{dx^j}.$$

14.3   The inverse metric tensor

The inverse of the metric tensor g_{ij} is written as g^{ij}. As a matrix, it is the inverse of g_{ij}, in shorthand,

$$g^{ij} = (g_{ij})^{-1},$$

or, in index notation,

$$g^{ij} g_{jk} = \delta^i_k.$$

This is the same as the *definition* of the inverse of a matrix:

$$\mathbf{H}^{-1}\mathbf{H} = \mathbf{I},$$

in which $\mathbf{I}$ is the unit matrix, the matrix presentation of Kronecker's tensor δ^i_k. Written out:

$$\overbrace{\begin{bmatrix} g^{11} & g^{12} \\ g^{21} & g^{22} \end{bmatrix}}^{g^{ij} \downarrow i, \rightarrow j} \overbrace{\begin{bmatrix} g_{11} & g_{12} \\ g_{21} & g_{22} \end{bmatrix}}^{g_{jk} \downarrow j, \rightarrow k} = \overbrace{\begin{bmatrix} 1 & 0 \\ 0 & 1 \end{bmatrix}}^{\delta^i_k \downarrow i, \rightarrow k}.$$

14.3.1   Raising or lowering tensor indices

An index of an arbitrary tensor may be *raised* or *lowered* by multiplying with the metric tensor g_{ij} or its inverse tensor g^{ij}:

$$T^i_j = g^{ik} T_{kj} = g_{jk} T^{ki}.$$

All forms T_{ij}, T^{ij}, T^i_j describe the same tensor, written in different ways. As a special case, $\delta^i_j = g_{jk}g^{ki} = g^{ik}g_{kj}$, i.e., the Kronecker delta tensor is the mixed version of the metric tensor and might be written as g^i_j. The delta way of writing has however become generally accepted.

14.3.2   The eigenvalues and eigenvectors of a tensor

The eigenvalue problem of a square tensor has the following forms:

$$\left(T^{ij} - \lambda g^{ij}\right) x_j = 0,$$
$$\left(T_{ij} - \lambda g_{ij}\right) x^j = 0,$$
$$\left(T^i_j - \lambda \delta^i_j\right) x^j = \left(T^i_j - \lambda \delta^i_j\right) x_i = 0.$$

All three forms are equivalent, as can be readily proven. For the eigenvectors, the following applies: $x_i = g_{ij} x^j$. If the tensor is *symmetric* (i.e., $T^i_j = T^j_i$), the eigenvalues λ are real valued and the eigenvectors mutually orthogonal: if x^i, y^i are different eigenvectors, then

$$g_{ij} x^i y^j = 0.$$

This can be read as the vanishing of a bilinear form or dot product, with notational variants

$$g_{ij} x^i y^j = x^i y_j = x_i y^j = \langle \mathbf{x} \cdot \mathbf{y} \rangle_{\mathbf{H}} = \mathbf{x}^{\mathsf{T}} \mathbf{H} \mathbf{y} = 0.$$

A tensor has as many eigenvalues as there are dimensions in the space, i.e., in the plane $\mathbb{R}^2$, tensors have two eigenvalues.

14.3.3   Visualization of a tensor

The quadratic form
$$T_{ij} x^i x^j = 1$$
defines an ellipsoid (in $\mathbb{R}^2$ an ellipse) that may be seen as the visualization of the tensor T_{ij}; for example, the inertial ellipsoid, the variance ellipsoid.

14.4   Parallellity of vectors and parallel transport

On a non-curved surface — and in the non-curved Euclidean space — the same-directedness of vectors, their *parallellity*, is not a problematic concept. If on the surface, or in the space, there exists a rectangular or Cartesian co-ordinate frame, then the vectorial values of a vector field at different places will be parallel, if all components are identical. The parallel transport of a vector, its transport without changing its direction, along a curve is similarly unproblematic.

On a curved surface, complications arise. However, for two vectors near each other we may define their parallellity by projecting the vectors onto the *tangent*

Figure 14.4: A parallel ruler for maritime use (Wikimedia Commons, Parallel ruler).

plane, and inspecting the parallellity of the projections. Now we execute the parallel transport along a curve in small steps such, that firstly we transport it parallelly a short distance d in the tangent plane of the starting point of the curve. After that, the vector is projected onto the tangent plane of the next point on the curve, which will be in a slightly different orientation. We repeat parallel transport and projection, and so forth, along the whole curve. This extended definition will work at least for those curved surfaces that are embedded in a non-curved Euclidean space.

Conceptually, one may think of parallel transport as involving a chain of tiny parallellograms: the defining property of a parallellogram is, that opposite sides are equal in length. This means that, if the metric, i.e., the distance between nearby points, is well defined, then so is parallel transport. A mechanical device for parallel transport of lines on a sea chart is the *parallel ruler* invented in 1584 by Fabrizio Mordente[10]. See Figure 14.4.

Of course, this is not just mathematics: gyroscopes are widely used precisely because they preserve their axis direction under parallel transport, e.g., Wikipedia, Gravity Probe B. And what is the old Foucault pendulum if not a device to observe the parallel transport of its direction of swing by the Earth's rotation?

14.5   The Christoffel symbols

The metric tensor is not yet the same thing as curvature. It does not even mean the curvature of co-ordinate lines: in order to study this, we need the *Christoffel*[11] *symbols*.

The Christoffel symbols describe what happens to a *vector* when it is being *transported parallelly* along a surface; more precisely, what happens with its components.

---

[10]Fabrizio Mordente (1532 – 1608) was an Italian mathematician.
[11]Elwin Bruno Christoffel (1829 – 1900) was a German mathematician and physicist.

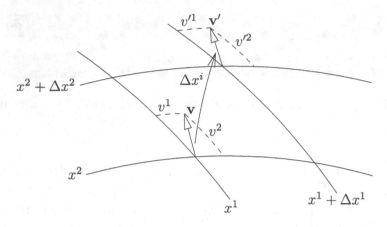

Figure 14.5: Christoffel symbols and parallel transport.

See Figure 14.5. When the vector $\mathbf{v}$, the components of which are v^i, is transported in a parallel fashion from one point to another over a distance Δx^i, its components will change by an amount $\Delta v^i = v'^{\,i} - v^i$, although "for the vector itself" $\mathbf{v}' = \mathbf{v}$. When both the vector being transported and the distance of transport are *small*, it can be shown that the change depends *linearly* both on the vector itself and on the direction of transport, as follows:

$$\Delta v^i = \Gamma^i_{jk} v^j \Delta x^k.$$

The Christoffel symbols Γ^i_{jk} *do not form a tensor*, unlike the metric tensor, from which the symbols may be computed. They describe the curvilinearity of the curvilinear co-ordinate frame, i.e., a property of the *co-ordinate frame*. A co-ordinate transformation that makes — at least locally, if not everywhere — the curvature of the co-ordinate curves vanish, will also set the elements of Γ^i_{jk} to zero in that point. For tensors, such a vanishing transformation cannot be found, not even locally.

For example, on the Earth's surface, in (φ, λ) co-ordinates, the curvature of the co-ordinate curves vanishes locally on the equator, and later we shall show that there, all Γ^i_{jk} vanish. Another example is from general relativity, in which the components of *acceleration* are Christoffel symbols: acceleration can be transformed away by going to a "falling along" reference frame. In Einstein's falling-elevator thought experiment, people inside the elevator are accelerating with respect to the Earth's surface, but are weightless (i.e., the acceleration is zero) in a reference frame connected to the elevator.

The Christoffel symbols can be calculated from the metric tensor; the equation is (complicated proof; see Appendix D on page 251):

$$\Gamma^i_{jk} = \tfrac{1}{2} g^{i\ell} \left(\frac{\partial g_{k\ell}}{\partial x^j} + \frac{\partial g_{\ell j}}{\partial x^k} - \frac{\partial g_{jk}}{\partial x^\ell} \right) \tag{14.6}$$

$(g^{ij} \overset{\text{def}}{=} (g_{ij})^{-1})$. From this it can be seen that the Christoffel symbols are like the first derivatives of the metric. In rectangular co-ordinates on a non-curved surface, g_{ij} is a constant and thus all Γ^i_{jk} vanish. Because g_{ij} is symmetric $(g_{ij} = g_{ji})$, we also obtain

$$\Gamma^i_{jk} = \Gamma^i_{kj}.$$

The Christoffel symbols are useful when we write the equation of the *geodesic*[12] in this formalism (see Equation 13.9 on page 196):

$$\frac{d}{ds}t^i + \Gamma^i_{jk}t^j t^k = 0.$$

In this, t^i is the *tangent vector* of the geodesic, $t^i = \dfrac{dx^i}{ds}$.

Example (1). The surface of a spherical Earth (φ, λ). Use the notation in which φ and λ are denoted by the index values 1 and 2 (for these symbols, Einstein's summation convention thus does *not* work!):

$$\frac{\partial g_{k\ell}}{\partial \varphi} = \frac{\partial}{\partial \varphi}\begin{bmatrix} R^2 & 0 \\ 0 & R^2 \cos^2 \varphi \end{bmatrix} = \begin{bmatrix} 0 & 0 \\ 0 & -R^2 \sin 2\varphi \end{bmatrix},$$

$$\frac{\partial g_{k\ell}}{\partial \lambda} = \frac{\partial}{\partial \lambda}\begin{bmatrix} R^2 & 0 \\ 0 & R^2 \cos^2 \varphi \end{bmatrix} = \begin{bmatrix} 0 & 0 \\ 0 & 0 \end{bmatrix},$$

so only

$$\frac{\partial g_{\lambda\lambda}}{\partial \varphi} = \frac{\partial g_{22}}{\partial \varphi} = -R^2 \sin 2\varphi$$

is different from zero. Remembering that

$$g^{i\ell} = (g_{i\ell})^{-1} = \begin{bmatrix} R^2 & 0 \\ 0 & R^2 \cos^2 \varphi \end{bmatrix}^{-1} = \begin{bmatrix} R^{-2} & 0 \\ 0 & R^{-2}\cos^{-2}\varphi \end{bmatrix},$$

the only elements different from zero are

$$\Gamma^\varphi_{\lambda\lambda} = \frac{1}{2}(g_{\varphi\varphi})^{-1}\left(-\frac{\partial g_{\lambda\lambda}}{\partial \varphi}\right) = \frac{1}{2}R^{-2}\cdot\left(+R^2 \sin 2\varphi\right) = +\frac{1}{2}\sin 2\varphi = \sin\varphi\cos\varphi,$$

$$\Gamma^\lambda_{\varphi\lambda} = \frac{1}{2}(g_{\lambda\lambda})^{-1}\left(\frac{\partial g_{\lambda\lambda}}{\partial \varphi}\right) = \frac{1}{2}\left(R^{-2}\cos^{-2}\varphi\right)\left(-R^2 \sin 2\varphi\right) = -\frac{\sin 2\varphi}{2\cos^2\varphi} = -\tan\varphi,$$

$$\Gamma^\lambda_{\lambda\varphi} = \frac{1}{2}(g_{\lambda\lambda})^{-1}\left(\frac{\partial g_{\lambda\lambda}}{\partial \varphi}\right) = -\tan\varphi.$$

$$\tag{14.7}$$

---

[12]In fact we write

$$\frac{Dt^i}{ds} \overset{\text{def}}{=} \frac{dt^i}{ds} + \Gamma^i_{jk}t^j t^k,$$

the so-called *absolute* or *covariant derivative*, which is a tensor, and

$$\frac{Dt^i}{ds} = 0$$

is then the equation for the geodesic, which thus applies in any curvilinear co-ordinate frame.

Note that, on the equator, $\varphi = 0$, so all $\Gamma^i_{jk} = 0$.

Example (2). Polar co-ordinates (ρ, θ) in the plane:

$$\frac{\partial g_{k\ell}}{\partial \rho} = \frac{\partial}{\partial \rho} \begin{bmatrix} 1 & 0 \\ 0 & \rho^2 \end{bmatrix} = \begin{bmatrix} 0 & 0 \\ 0 & 2\rho \end{bmatrix}, \quad \frac{\partial g_{k\ell}}{\partial \theta} = \frac{\partial}{\partial \varphi} \begin{bmatrix} 1 & 0 \\ 0 & \rho^2 \end{bmatrix} = \begin{bmatrix} 0 & 0 \\ 0 & 0 \end{bmatrix}.$$

The only element different from zero is

$$\frac{\partial g_{22}}{\partial \rho} = \frac{\partial g_{\theta\theta}}{\partial \rho} = 2\rho,$$

yielding

$$\Gamma^\rho_{\theta\theta} = \frac{1}{2} (g_{\rho\rho})^{-1} \left(-\frac{\partial g_{\theta\theta}}{\partial \rho} \right) = \frac{1}{2} \cdot -2\rho = -\rho,$$

$$\Gamma^\theta_{\rho\theta} = \Gamma^\theta_{\theta\rho} = \frac{1}{2} (g_{\theta\theta})^{-1} \left(+\frac{\partial g_{\theta\theta}}{\partial \rho} \right) = \frac{1}{2} \cdot \frac{1}{\rho^2} \cdot 2\rho = +\frac{1}{\rho}.$$

14.6   Alternative equations for the geodesic

Here we give alternative equations for integrating the geodesic on the sphere (generalization to the *ellipsoid of revolution* is complicated but possible):

$$\frac{d\xi}{ds} + \eta^2 \sin \varphi \cos \varphi = 0,$$

$$\frac{d\eta}{ds} - 2\eta\xi \tan \varphi = 0.$$

Here we have used the expressions 14.7 for the elements of Γ^i_{jk} in (φ, λ) co-ordinates on the sphere.

Simultaneous integration of these differential equations now yields, as functions of s,

$$\begin{bmatrix} \xi \\ \eta \end{bmatrix} = \begin{bmatrix} \cos A / R \\ \sin A / R \cos \varphi \end{bmatrix}$$

from which A and φ may be calculated.

We add to this system the equations defining the tangent

$$\frac{d\varphi}{ds} = \xi = t^1,$$

$$\frac{d\lambda}{ds} = \eta = t^2.$$

In this way, also λ comes along.

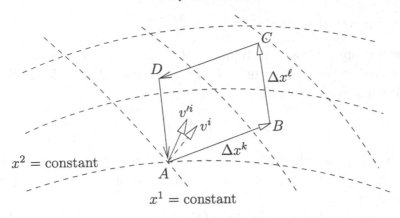

Figure 14.6: The curvature tensor and parallel transport around a closed parallellogram path.

This approach may seem unnecessarily complicated. Its significant advantage from a theoretical viewpoint is that it works in *all* curvilinear co-ordinate systems on the surface under consideration, also, e.g., in stereographic projection co-ordinates used for polar areas. A requirement is, however, that the *metric tensor* of the co-ordinate curves is first written out.

The *length* of the tangent vector $\begin{bmatrix} \xi & \eta \end{bmatrix}^{\mathsf{T}}$ is computed as follows:

$$ds^2 = g_{ij}t^i t^j = R^2 \left(\frac{\cos A}{R} \right)^2 + R^2 \cos^2 \varphi \left(\frac{\sin A}{R \cos \varphi} \right)^2 = 1,$$

remembering that

$$g_{ij} = \begin{bmatrix} R^2 & 0 \\ 0 & R^2 \cos^2 \varphi \end{bmatrix}$$

on the surface of the sphere. The length of the tangent vector must always be 1. This requirement is fulfilled if s is a parametrisation of the curve by arc length.

14.7 The curvature tensor

Curvature is described in a rather complicated way — by means of *parallel transport* around a *closed path* — the edge of a small "patch," more precisely, of a *parallellogram.*

As seen in Figure 14.6, we obtain for the change in the components of vector v^i, $\Delta v^i \stackrel{\text{def}}{=} (v')^i - v^i$, which depends on

1. the vectors of the two sides of the parallellogram, consisting of co-ordinate differences, $\Delta x^k = x_B^k - x_A^k$ and $\Delta x^\ell = x_C^\ell - x_B^\ell$. These vectors must be of different directions, or this will not work.

2. the transported vector v^i itself.

These observations are cast in the form of an equation as follows, all the time assuming that the parallellogram and the quantities Δ are all small, and thus, that this linearization is valid:

$$\Delta v^i = R^i_{jk\ell} v^j \Delta x^k \Delta x^\ell. \tag{14.8}$$

This creature $R^i_{jk\ell}$ is called the *Riemann curvature tensor*, or sometimes the Riemann–Christoffel tensor. It has, in two-dimensional space (i.e., on a surface), $2^4 = 16$ elements.

The equation for the computation of the Riemann tensor from the Christoffel symbols is derived in Appendix E on page 253:

$$R^i_{jk\ell} = \frac{\partial}{\partial x^k}\Gamma^i_{j\ell} - \frac{\partial}{\partial x^\ell}\Gamma^i_{jk} + \Gamma^i_{km}\Gamma^m_{j\ell} - \Gamma^i_{\ell m}\Gamma^m_{jk}. \tag{14.9}$$

Here, immediately the following antisymmetry is seen:

$$R^i_{jk\ell} = -R^i_{j\ell k}.$$

Many other similar symmetries can be found. The number of independent components of the Riemann tensor is actually quite small.

Example (1). The surface (φ, λ) of a spherical Earth. Based on the antisymmetry property, we may say that only the elements for which $k \neq \ell$ can be different from zero. Additionally, the Christoffel symbols depend only on φ. In this way we obtain the auxiliary terms:

$$\frac{\partial}{\partial \varphi}\Gamma^\varphi_{\lambda\lambda} = \frac{\partial}{\partial \varphi}\left(\tfrac{1}{2}\sin 2\varphi\right) = +\cos 2\varphi,$$

$$\frac{\partial}{\partial \varphi}\Gamma^\lambda_{\varphi\lambda} = \frac{\partial}{\partial \varphi}\Gamma^\lambda_{\lambda\varphi} = \frac{\partial}{\partial \varphi}\left(-\tan\varphi\right) = -\frac{1}{\cos^2\varphi},$$

the others being zero.

The terms $\Gamma^i_{km}\Gamma^m_{j\ell}$ are obtained as follows:

$$\Gamma^\varphi_{\lambda\lambda}\Gamma^\lambda_{\varphi\lambda} = \Gamma^\lambda_{\lambda\varphi}\Gamma^\varphi_{\lambda\lambda} = \Gamma^\varphi_{\lambda\lambda}\Gamma^\lambda_{\lambda\varphi} = -\sin^2\varphi,$$

$$\Gamma^\lambda_{\varphi\lambda}\Gamma^\lambda_{\lambda\varphi} = \Gamma^\lambda_{\varphi\lambda}\Gamma^\lambda_{\varphi\lambda} = +\tan^2\varphi.$$

By combining, we obtain

$$R^\varphi_{\varphi\varphi\lambda} = -R^\varphi_{\varphi\lambda\varphi} = \frac{\partial}{\partial\varphi}\Gamma^\varphi_{\varphi\lambda} - \frac{\partial}{\partial\lambda}\Gamma^\varphi_{\varphi\varphi} + \Gamma^\varphi_{\varphi m}\Gamma^m_{\varphi\lambda} - \Gamma^\varphi_{\lambda m}\Gamma^m_{\varphi\varphi} = 0;$$

$$R^\lambda_{\varphi\varphi\lambda} = -R^\lambda_{\varphi\lambda\varphi} = \frac{\partial}{\partial\varphi}\Gamma^\lambda_{\varphi\lambda} - \frac{\partial}{\partial\lambda}\Gamma^\lambda_{\varphi\varphi} + \Gamma^\lambda_{\varphi m}\Gamma^m_{\varphi\lambda} - \Gamma^\lambda_{\lambda m}\Gamma^m_{\varphi\varphi} =$$

$$= -\frac{1}{\cos^2\varphi} + \tan^2\varphi = -1;$$

$$R^{\varphi}_{\lambda\varphi\lambda} = -R^{\varphi}_{\lambda\lambda\varphi} = \frac{\partial}{\partial\varphi}\Gamma^{\varphi}_{\lambda\lambda} - \frac{\partial}{\partial\lambda}\Gamma^{\varphi}_{\lambda\varphi} + \Gamma^{\varphi}_{\varphi m}\Gamma^{m}_{\lambda\lambda} - \Gamma^{\varphi}_{\lambda m}\Gamma^{m}_{\lambda\varphi} =$$

$$= \cos 2\varphi + \sin^2\varphi = \cos^2\varphi;$$

$$R^{\lambda}_{\lambda\varphi\lambda} = -R^{\lambda}_{\lambda\lambda\varphi} = \frac{\partial}{\partial\varphi}\Gamma^{\lambda}_{\lambda\lambda} - \frac{\partial}{\partial\lambda}\Gamma^{\lambda}_{\lambda\varphi} + \Gamma^{\lambda}_{\varphi m}\Gamma^{m}_{\lambda\lambda} - \Gamma^{\lambda}_{\lambda m}\Gamma^{m}_{\lambda\varphi} = 0.$$

Example (2). Polar co-ordinates in the (ρ, θ) plane:

$$\frac{\partial}{\partial\rho}\Gamma^{\rho}_{\theta\theta} = -1,$$

$$\frac{\partial}{\partial\rho}\Gamma^{\theta}_{\rho\theta} = \frac{\partial}{\partial\rho}\Gamma^{\theta}_{\theta\rho} = -\frac{1}{\rho^2},$$

the others again being zero. Then:

$$\Gamma^{\rho}_{\theta\theta}\Gamma^{\theta}_{\rho\theta} = \Gamma^{\theta}_{\theta\rho}\Gamma^{\rho}_{\theta\theta} = \Gamma^{\rho}_{\theta\theta}\Gamma^{\theta}_{\theta\rho} = -1,$$

$$\Gamma^{\theta}_{\rho\theta}\Gamma^{\theta}_{\theta\rho} = \Gamma^{\theta}_{\rho\theta}\Gamma^{\theta}_{\rho\theta} = +\frac{1}{\rho^2}.$$

After this:

$$R^{\rho}_{\rho\rho\theta} = -R^{\theta}_{\rho\theta\rho} = 0;$$

$$R^{\theta}_{\rho\rho\theta} = -R^{\theta}_{\rho\theta\rho} = \frac{\partial}{\partial\rho}\Gamma^{\theta}_{\rho\theta} - \frac{\partial}{\partial\theta}\Gamma^{\theta}_{\rho\rho} + \Gamma^{\theta}_{\rho m}\Gamma^{m}_{\rho\theta} - \Gamma^{\theta}_{\theta m}\Gamma^{m}_{\rho\rho} = -\frac{1}{\rho^2} + \frac{1}{\rho^2} = 0;$$

$$R^{\rho}_{\theta\rho\theta} -- -R^{\rho}_{\theta\theta\rho} = \frac{\partial}{\partial\rho}\Gamma^{\rho}_{\theta\theta} - \frac{\partial}{\partial\theta}\Gamma^{\rho}_{\theta\rho} + \Gamma^{\rho}_{\rho m}\Gamma^{m}_{\theta\theta} - \Gamma^{\rho}_{\theta m}\Gamma^{m}_{\theta\rho} = -1 + 1 = 0;$$

$$R^{\theta}_{\theta\rho\theta} = -R^{\theta}_{\theta\theta\rho} = 0.$$

So the *whole Riemann tensor vanishes* as it should because the surface is not curved.

From the Riemann tensor is obtained the smaller Ricci[13] tensor in the following way:

$$R_{jk} = R^{i}_{jik} = \sum_{i=1}^{2} R^{i}_{jik}. \qquad (14.10)$$

This is a symmetric tensor, $R_{ij} = R_{ji}$.

Example (1). Continue with the computation of R_{ij} for the spherical surface:

$$R_{\varphi\varphi} = R^{\varphi}_{\varphi\varphi\varphi} + R^{\lambda}_{\varphi\lambda\varphi} = +1;$$

$$R_{\varphi\lambda} = R^{\varphi}_{\varphi\varphi\lambda} + R^{\lambda}_{\varphi\lambda\lambda} = 0 + 0 = 0;$$

$$R_{\lambda\lambda} = R^{\lambda}_{\lambda\lambda\lambda} + R^{\varphi}_{\lambda\varphi\lambda} = \cos^2\varphi.$$

Example (2). For polar co-ordinates in the plane, all $R_{ij} = 0$.

---

[13]Gregorio Ricci-Curbastro (1853–1925) was an Italian mathematician and inventor of tensor calculus.

We may continue this process to produce the *curvature scalar* or Ricci scalar[14]:

$$\mathcal{R} = g^{ij} R_{ji} = \sum_{j=1}^{2} \sum_{i=1}^{2} (g_{ij})^{-1} R_{ji}.$$

Example (1). Spherical surface:

$$R_k^i = g^{ij} R_{jk} = \begin{bmatrix} \dfrac{1}{R^2} & 0 \\ 0 & \dfrac{1}{R^2 \cos^2 \varphi} \end{bmatrix} \begin{bmatrix} 1 & 0 \\ 0 & +\cos^2 \varphi \end{bmatrix} = \begin{bmatrix} R^{-2} & 0 \\ 0 & R^{-2} \end{bmatrix},$$

so[15]

$$\mathcal{R} = \sum_i R_i^i = \frac{1}{R^2} + \frac{\cos^2 \varphi}{R^2 \cos^2 \varphi} = \frac{2}{R^2}.$$

Example (2). In polar co-ordinates, $R_{ij} = 0$ and thus $\mathcal{R} = 0$.

More generally applies (without proof):

$$\mathcal{R} = 2K = 2\kappa_1 \kappa_2 = \frac{2}{R_1 R_2}, \tag{14.11}$$

twice the *total curvature value* K of Gauss, the inverse of the product of the principal radii of curvature. See also Equation 7.1.

Exercises

Exercise 14−1: Geometry of the torus

A *torus* (doughnut) can be described parametrically in the following way:

$$\mathbf{x} = \begin{bmatrix} (a + b \cos u) \cos v \\ (a + b \cos u) \sin v \\ b \sin u \end{bmatrix}$$

in which a is the "major radius" and b the "minor radius." See Figure 14.7.

a. Calculate the metric tensor $g_{ij}(u, v)$ of the surface of the torus.

b. Calculate the Christoffel symbols $\Gamma^i_{jk}(u, v)$ of the torus.

c. Calculate the curvature tensor $R^i_{jk\ell}$ of the torus (*difficult*).

---

[14]This is actually the *trace* of the tensor R_{ij}, more precisely that of $R^i_j \overset{\text{def}}{=} g^{ik} R_{kj}$, i.e., R^i_i, see above. Let this also be an example of how, in general curvilinear co-ordinates, an index may be "raised" from covariant to contravariant, or "lowered," using the metric tensor g_{ij} or its inverse g^{ij}. In rectangular co-ordinates, these are of course the unit matrices, and the difference between super- and subscripts is without significance.

[15]Do not let the use of similar symbols confuse you, R vs. $\mathcal{R}$, for the radius of the Earth and the curvature scalar.

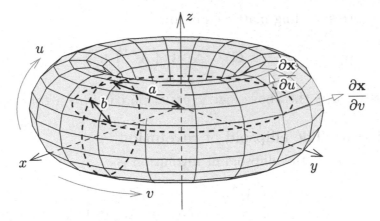

Figure 14.7: The geometry of the torus.

Exercise 14 – 2: Transformation of the metric tensor

Let the co-ordinates (u, v) be defined as follows:

$$u = \tfrac{1}{2}\left(x^2 - y^2\right),$$

$$v = xy.$$

a. Calculate the co-ordinate transformation matrix

$$du^{i'} = \alpha_i^{i'} du^i,$$

in which

$$du^{i'} = \begin{bmatrix} du \\ dv \end{bmatrix}, \quad du^i = \begin{bmatrix} dx \\ dy \end{bmatrix}$$

In other words, write

$$du = \alpha_1^1 dx + \alpha_2^1 dy,$$
$$dv = \alpha_1^2 dx + \alpha_2^2 dy$$

and give the coefficients.

b. What is the inverse matrix $\alpha_{i'}^i$?

c. If the metric of the (x, y) co-ordinates is

$$g_{ij} = \begin{bmatrix} 1 & 0 \\ 0 & 1 \end{bmatrix}$$

(Euclidean metric), calculate the metric of the (u, v) co-ordinates using the *tensor transformation formula*

$$g_{i'j'} = \sum_{i,j} \alpha_{i'}^i \alpha_{j'}^j g_{ij},$$

or the corresponding matric equation

$$\mathbf{G}' = \mathbf{A}^{-1}\mathbf{G}\left(\mathbf{A}^{-1}\right)^{\mathsf{T}}.$$

d. Reflect on *why* the above matric equation contains a *transposition sign*.

Exercise 14−3: The ellipsoid of revolution

The metric tensor on the ellipsoid of revolution is

$$g_{ij} = \begin{bmatrix} M^2\left(\varphi\right) & 0 \\ 0 & \left(N\left(\varphi\right)\cos\varphi\right)^2 \end{bmatrix}.$$

a. Calculate the Christoffel symbols Γ^i_{jk}. Equation:

$$\Gamma^i_{jk} = \tfrac{1}{2}g^{i\ell}\left(\frac{\partial g_{k\ell}}{\partial x^j} + \frac{\partial g_{\ell j}}{\partial x^k} - \frac{\partial g_{jk}}{\partial x^\ell}\right).$$

(So, calculate $\Gamma^\varphi_{\varphi\varphi}, \Gamma^\varphi_{\varphi\lambda}$, etc., altogether $2^3 = 8$ elements). For this you need Appendix C on page 249.

b. Check that the result is compatible, in the limit $e \to 0$, with the earlier derived results for the sphere.

c. Using the above results, derive from the general form for the tangent vector of the geodesic:

$$\frac{d}{ds}t^i + \Gamma^i_{jk}t^j t^k = 0$$

the form suitable for the ellipsoid of resolution (i.e., $\frac{dt^\varphi}{ds} = \cdots$, $\frac{dt^\lambda}{ds} = \cdots$).

Exercise 14−4: The Einstein tensor

The *Einstein tensor* may be defined as

$$G_{ij} \stackrel{\text{def}}{=} R_{ij} - \frac{2}{n}g_{ij}\mathcal{R},$$

with n the dimensionality of the space.

1. Derive, for a spherical surface, i.e., $n = 2$, of radius R, this tensor in terms of the total curvature $K = R^{-2}$.

2. Derive the *Einstein scalar*, also in terms of the total curvature.

$$\mathcal{G} \stackrel{\text{def}}{=} g^{ij}G_{ij}.$$

3. The Einstein field equations in the general theory of relativity are

$$G_{\mu\nu} + \Lambda g_{\mu\nu} = \frac{8\pi G}{c^4} T_{\mu\nu},$$

with $T_{\mu\nu}$ the energy-momentum tensor describing masses and mass flows in spacetime, and Λ the cosmological constant ("dark energy"). G is the universal gravitational constant, c the speed of light in vacuum.

We may postulate in two dimensions a similar, scalar field equation of the form

$$\mathcal{G} = -4\pi G \rho$$

with ρ a scalar mass density. With a field equation of this form, derive the *total mass*

$$M \stackrel{\text{def}}{=} 4\pi \rho R^2$$

of this toy universe.

Chapter 15

Map projections in light of surface theory

Map projections are needed because mapping the curved surface of the Earth to a plane is impossible without error at least for an extended area. In this chapter, we look at the deformations in *scale* caused by the map projection, using the tools developed in the previous chapter.

15.1   The Gauss total curvature and spherical excess

Unlike the eigenvalues of the second fundamental form of Gauss, $\beta^i_k = g^{ij}\beta_{jk}$ (Section 13.4 on page 188), the eigenvalues of the tensor R^i_j are *not* connected to the principal curvatures of the surface, κ_1 and κ_2. The *radii* of curvature exist only in the three-dimensional space surrounding the Earth's surface, into which the surface is embedded. R^i_j, on the other hand, is an *intrinsic* property of the surface.

In the *two-dimensional case*, both the R_{ij} tensor and the $R^i_{jk\ell}$ tensor have *only one essentially independent element*, which is associated with the Gaussian total curvature $K = \kappa_1\kappa_2$.

Stated again differently, the curvature of a surface may be characterized by *one number K*, which may be calculated from *only the metric intrinsic to the surface* — by well-known operations like differentiation, multiplication, addition — without knowing anything about the properties of the space surrounding the surface or the way in which the surface is embedded in it[1]. This conceptually important result observed by Gauss is known by the name *theorema egregium*, "remarkable theorem," Bhatia (2014).

This may be demonstrated as follows:

1. in Equation 14.8 on page 220 $\Delta x^k \Delta x^\ell$ describes the sides of a small parallellogram, around which a vector v^i is transported in a parallel fashion. In two dimensions, only two independent side vectors $\Delta x^k, \Delta x^\ell$

---

[1]Physical interpretation: if the surface has been made, e.g., from a thin plastic foil, it may be bent *without stretching*, without the total curvature changing in any point. A plane map may be rolled up into a cylinder without the map projection changing. It can, however, *not* be rolled up into a globe without stretching!

may be chosen. And then $\Delta x^k \Delta x^\ell$ and $\Delta x^\ell \Delta x^k$ represent two possible directions of traversal, one clockwise and one counterclockwise. The corresponding elements $R^i_{jk\ell} = -R^i_{j\ell k}$ are thus essentially the same.

2. In the same equation, v^j and Δv^i are mutually perpendicular. Δv^i represents a small *rotation* of the vector v^j. On a surface, in two dimensions, the rotation is described by *one angle*. Because the angle between two parallelly transported vectors does not change, we may infer that this rotation angle is the same for all possible vectors v^i. Again, we find only one independent parameter. See Figure 15.1.

In other words, when given the parallellogram $dx^1 dx^2$, the expression

$$R^i_{j12}\sqrt{g}\, dx^1 dx^2 = -R^i_{j21}\sqrt{g}\, dx^2 dx^1,$$

is the 2×2 matrix describing the rotation of the vector

$$\begin{bmatrix} \cos\theta & -\sin\theta \\ \sin\theta & \cos\theta \end{bmatrix} - \begin{bmatrix} 1 & 0 \\ 0 & 1 \end{bmatrix} \approx \begin{bmatrix} 0 & -\theta \\ \theta & 0 \end{bmatrix}$$

in which there is only one free parameter. In this, $g = \det g_{ij} = g_{11}g_{22} - g_{12}g_{21}$, the determinant of the metric[2].

However, for higher dimensionalities, the number of independent elements of the Riemann and Ricci tensors is larger. This case is interesting from the perspective of the general theory of relativity (four dimensions!), though not from the perspective of geodesy.

We may note here that the *spherical excess* discussed already in Section 7.1 on page 77 is a special case of the change in direction of a vector transported parallelly around. The spherical excess is the small change in direction of a vector when it is transported around a closed *triangle*!

---

[2]This is in order to account for the surface area of an $x^1 x^2$ unit cell not being unity. The surface area is

$$\left\| \frac{\partial \mathbf{x}}{\partial x^1} \times \frac{\partial \mathbf{x}}{\partial x^2} \right\| = \left\| \frac{\partial \mathbf{x}}{\partial x^1} \right\| \left\| \frac{\partial \mathbf{x}}{\partial x^2} \right\| \sin\angle\left(\frac{\partial \mathbf{x}}{\partial x^1}, \frac{\partial \mathbf{x}}{\partial x^2} \right).$$

Let us call for brevity $\mathbf{x}_u \overset{\text{def}}{=} \dfrac{\partial \mathbf{x}}{\partial x_1}$ and $\mathbf{x}_v \overset{\text{def}}{=} \dfrac{\partial \mathbf{x}}{\partial x_2}$. Then, the above reads

$$\|\mathbf{x}_u \times \mathbf{x}_v\| = \|\mathbf{x}_u\| \, \|\mathbf{x}_v\| \sin\angle\,(\mathbf{x}_u, \mathbf{x}_v).$$

Its square is

$$\|\mathbf{x}_u\|^2 \, \|\mathbf{x}_v\|^2 \sin^2\angle\,(\mathbf{x}_u, \mathbf{x}_v) = \|\mathbf{x}_u\|^2 \, \|\mathbf{x}_v\|^2 - \|\mathbf{x}_u\|^2 \, \|\mathbf{x}_v\|^2 \cos^2\angle\,(\mathbf{x}_u, \mathbf{x}_v) =$$
$$= \langle\mathbf{x}_u \cdot \mathbf{x}_u\rangle \langle\mathbf{x}_v \cdot \mathbf{x}_v\rangle - \langle\mathbf{x}_u \cdot \mathbf{x}_v\rangle^2 =$$
$$= g_{11}g_{22} - g_{12}g_{21} = \det g_{ij}.$$

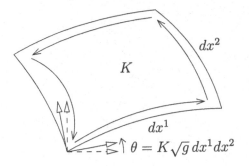

$$\theta = K\sqrt{g}\,dx^1 dx^2$$

Figure 15.1: The curvature of a two-dimensional surface is characterized by only one parameter: K.

When we transport a vector around a larger surface area, it is the same as if it were transported successively around the small patches that make up the surface area. This is presented in Figure 15.2. In every small patch $dx^1 dx^2$, of surface area $dS = \sqrt{g}\,dx^1 dx^2$, the change in direction of the vector has the magnitude $K\,dS$, in which K is the Gaussian total curvature at the patch.

From this, one obtains by generalization the following integral equation (the Gauss–Bonnet[3] theorem, *theorema elegantissimum*, for a triangle):

$$\varepsilon = \iint\limits_{\Delta} K\left(\varphi, \lambda\right) dS,$$

in which K is a function of place.

On the reference ellipsoid $K = (MN)^{-1}$, $g = \det g_{ij} = M^2 \cdot N^2 \cos^2 \varphi$ and thus

$$\varepsilon = \iint\limits_{\Delta} \frac{1}{MN}\,dS = \iint\limits_{\Delta} \frac{1}{MN} MN \cos\varphi\,d\varphi d\lambda = \iint\limits_{\Delta} d\sigma = \sigma_{\Delta},$$

the area of the corresponding triangle on the surface of the *unit sphere*[4]. From this follows that

---

[3]Pierre Ossian Bonnet (1819 – 1892) was a French mathematician.

[4]This *does not apply exactly*: the triangle symbols under the ellipsoidal integral $d\varphi d\lambda$ and under the spherical integral $d\sigma$ do not exactly match. On the reference ellipsoid, the triangle consists of geodesics, while on the unit sphere it consists of great-circle segments. However, the ellipsoidal geodesics *do not map* precisely onto the great circles!

This may already be suspected from the observation that a long geodesic does not generally close upon itself — unlike a great circle around a sphere.

Furthermore, the *angles* of the ellipsoidal and spherical triangle are not individually of the same size; their *sums* (and with that, their spherical excesses) however are.

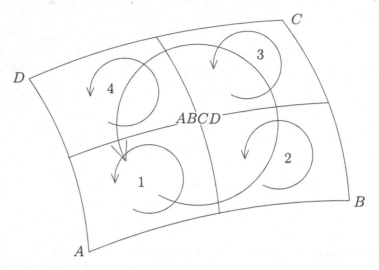

Figure 15.2: Transport of a vector around a larger surface area. $\theta_{ABCD} = \theta_1 + \theta_2 + \theta_3 + \theta_4$.

> The spherical excess on the reference ellipsoid depends only on the directions (φ_i, λ_i), $i = 1, 2, 3$ of the ellipsoidal normals in the three vertices of the triangle, *and not of the figure of the ellipsoid*. Shorter: *spherical excesses are computed on the sphere, even if the other computations are done on the ellipsoid.*

It suffices if the geodetic co-ordinates of the vertices, which characterize the ellipsoidal normal, are known.

A fast geometric way of computing the excess of a spherical triangle is the following. Let the vertices be

$$\mathbf{x}_i = \begin{bmatrix} \cos \varphi_i \cos \lambda_i \\ \cos \varphi_i \sin \lambda_i \\ \sin \varphi_i \end{bmatrix}, \, i = 1, 2, 3.$$

Use the *polarization method* (see Section 7.6 on page 84) in three dimensions to find the poles[5] of the triangle's sides:

$$\mathbf{y}_1 = \frac{\langle \mathbf{x}_2 \times \mathbf{x}_3 \rangle}{\| \langle \mathbf{x}_2 \times \mathbf{x}_3 \rangle \|}, \qquad \mathbf{y}_2 = \frac{\langle \mathbf{x}_3 \times \mathbf{x}_1 \rangle}{\| \langle \mathbf{x}_3 \times \mathbf{x}_1 \rangle \|}, \qquad \mathbf{y}_3 = \frac{\langle \mathbf{x}_1 \times \mathbf{x}_2 \rangle}{\| \langle \mathbf{x}_1 \times \mathbf{x}_2 \rangle \|}.$$

---

[5]So we exploit that, if we give to some side of the triangle the role of equator, the corresponding pole will be at a distance $\frac{\pi}{2}$, i.e., 90°, from it, and from the triangle's vertices on it. Then, the cross-product of the unit vectors pointing to these vertices will give the direction of the pole.

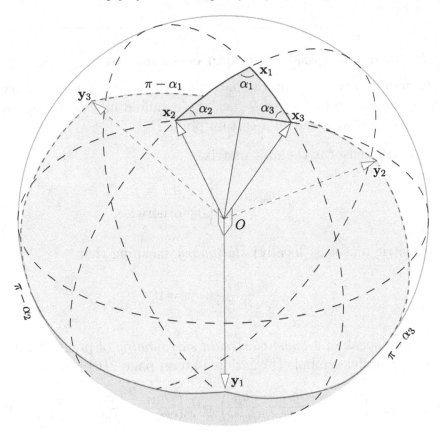

Figure 15.3: Using polarization to calculate the spherical excess. The angles of the triangle are calculated from the unit vectors $\mathbf{y}_i$ pointing to the vertices of the polarization triangle, which differ in direction by the supplements of these angles.

See Figure 15.3. Now the distances between the poles are obtained from the following numerically strong equations for the angles of the spherical triangle (more precisely, π minus the angles in question):

$$\alpha_1 = \pi - 2\arctan\frac{\|\mathbf{y}_2 - \mathbf{y}_3\|}{\|\mathbf{y}_2 + \mathbf{y}_3\|},$$

$$\alpha_2 = \pi - 2\arctan\frac{\|\mathbf{y}_1 - \mathbf{y}_3\|}{\|\mathbf{y}_1 + \mathbf{y}_3\|},$$

$$\alpha_3 = \pi - 2\arctan\frac{\|\mathbf{y}_1 - \mathbf{y}_2\|}{\|\mathbf{y}_1 + \mathbf{y}_2\|}.$$

Here, α_i is the angle of vertex i. The spherical excess is now

$$\varepsilon = \sum_{i=1}^{3} \alpha_i - \pi.$$

15.2   Curvature in quasi-Cartesian co-ordinates

Let us mention without proof that, on a Riemann manifold, curvilinear co-ordinates may be *transformed* — i.e., the manifold may be re-parametrized — always in such a way that in a certain point P

1. the metric tensor is the unit matrix,

$$g_{ij} = g^{ij} = \begin{cases} 1 & \text{if } i = j, \\ 0 & \text{otherwise,} \end{cases}$$

2. the metric tensor is (locally) *stationary*, meaning that

$$\left. \frac{\partial g_{ij}}{\partial x^k} \right|_{x^k = x_P^k} = 0.$$

In this case we speak of a *quasi-Cartesian surrounding* of point P. See Figure 15.4. The Christoffel symbols (Equation 14.6 on page 216) are now

$$\Gamma_{jk}^i = \tfrac{1}{2} g^{i\ell} \left(\frac{\partial g_{k\ell}}{\partial x^j} + \frac{\partial g_{\ell j}}{\partial x^k} - \frac{\partial g_{jk}}{\partial x^\ell} \right) = \tfrac{1}{2} \left(\frac{\partial g_{ki}}{\partial x^j} + \frac{\partial g_{ij}}{\partial x^k} - \frac{\partial g_{jk}}{\partial x^i} \right) \approx 0$$

based on the above assumption of stationarity of the metric. Of the Riemann curvature tensor (Equation 14.9 on page 220) the two last terms vanish:

$$\begin{aligned}
R_{jk\ell}^i &= \frac{\partial}{\partial x^k} \Gamma_{j\ell}^i - \frac{\partial}{\partial x^\ell} \Gamma_{jk}^i = \\
&= \tfrac{1}{2} \frac{\partial}{\partial x^k} \left(\frac{\partial g_{\ell i}}{\partial x^j} + \frac{\partial g_{ij}}{\partial x^\ell} - \frac{\partial g_{j\ell}}{\partial x^i} \right) - \tfrac{1}{2} \frac{\partial}{\partial x^\ell} \left(\frac{\partial g_{ki}}{\partial x^j} + \frac{\partial g_{ij}}{\partial x^k} - \frac{\partial g_{jk}}{\partial x^i} \right) = \\
&= \tfrac{1}{2} \frac{\partial}{\partial x^j} \left(\frac{\partial g_{\ell i}}{\partial x^k} - \frac{\partial g_{ki}}{\partial x^\ell} \right) - \tfrac{1}{2} \frac{\partial}{\partial x^i} \left(\frac{\partial g_{j\ell}}{\partial x^k} - \frac{\partial g_{jk}}{\partial x^\ell} \right).
\end{aligned}$$

If we now compute the Ricci tensor (Equation 14.10 on page 221):

$$\begin{aligned}
R_{j\ell} = R_{ji\ell}^i &= -\tfrac{1}{2} \sum_i \frac{\partial^2 g_{j\ell}}{(\partial x^i)^2} + \tfrac{1}{2} \sum_i \left(\frac{\partial^2 g_{\ell i}}{\partial x^i \partial x^j} - \frac{\partial^2 g_{ii}}{\partial x^\ell \partial x^j} + \frac{\partial^2 g_{ji}}{\partial x^\ell \partial x^i} \right) = \\
&= -\tfrac{1}{2} \Delta g_{j\ell} + \tfrac{1}{2} \sum_i \left(\frac{\partial^2 g_{\ell i}}{\partial x^i \partial x^j} + \frac{\partial^2 g_{ij}}{\partial x^i \partial x^\ell} - \frac{\partial^2 g_{ii}}{\partial x^\ell \partial x^j} \right).
\end{aligned}$$

This already looks a lot more symmetric. The symbol Δ here means the *Laplace operator*

$$\Delta = \sum_i \frac{\partial^2}{(\partial x^i)^2}.$$

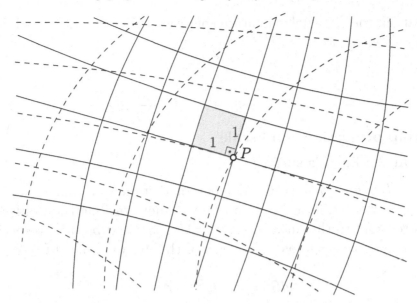

Figure 15.4: A quasi-Cartesian co-ordinate frame in a small surroundings of point P is obtained by a co-ordinate transformation (re-parametrization) from the general curvilinear parametrization or co-ordinate frame (dashed lines).

Finally, we compute still the curvature scalar

$$\mathcal{R} = g^{jk} R_{kj} = \sum_j R_{jj} =$$

$$= -\tfrac{1}{2} \sum_j \Delta g_{jj} + \sum_i \sum_j \frac{\partial^2 g_{ij}}{\partial x^i \partial x^j} - \tfrac{1}{2} \sum_i \sum_j \frac{\partial^2 g_{ii}}{(\partial x^j)^2} =$$

$$= -\tfrac{1}{2} \sum_j \Delta g_{jj} + \sum_i \sum_j \frac{\partial^2 g_{ij}}{\partial x^i \partial x^j} - \tfrac{1}{2} \sum_i \Delta g_{ii} =$$

$$= -\sum_i \Delta g_{ii} + \sum_i \sum_j \frac{\partial^2 g_{ij}}{\partial x^i \partial x^j}.$$

In the special case where the co-ordinate curves are (everywhere, not just in point P) *orthogonal*, we obtain $g_{ij} = 0$ if $i \neq j$ with its derivatives of place, i.e.,

$$\mathcal{R} = -\sum_i \Delta g_{ii} + \sum_i \frac{\partial^2 g_{ii}}{(\partial x^i)^2} =$$

$$= -\left(\frac{\partial^2}{(\partial x^1)^2} + \frac{\partial^2}{(\partial x^2)^2} \right) (g_{11} + g_{22}) + \left(\frac{\partial^2 g_{11}}{(\partial x^1)^2} + \frac{\partial^2 g_{22}}{(\partial x^2)^2} \right)$$

$$= -\left(\frac{\partial^2 g_{11}}{(\partial x^1)^2} + \frac{\partial^2 g_{11}}{(\partial x^2)^2} + \frac{\partial^2 g_{22}}{(\partial x^1)^2} + \frac{\partial^2 g_{22}}{(\partial x^2)^2} \right) + \left(\frac{\partial^2 g_{11}}{(\partial x^1)^2} + \frac{\partial^2 g_{22}}{(\partial x^2)^2} \right) =$$

$$= -\left(\frac{\partial^2 g_{11}}{(\partial x^2)^2} + \frac{\partial^2 g_{22}}{(\partial x^1)^2} \right). \tag{15.1}$$

This equation, the Chovitz[6] curvature condition (Chovitz, 1952), will be useful in the following section.

15.3   Map projections and scale

15.3.1   On the Earth's surface

On the Earth's surface, an element of distance dS may consist of an element of latitude difference $d\varphi$ and an element of longitude difference $d\lambda$. These correspond to the linear distances $M(\varphi)d\varphi$ and $p(\varphi)d\lambda$, respectively.

According to Pythagoras, the length of the diagonal of a $(d\varphi, d\lambda)$ postage stamp is

$$dS^2 = M^2 d\varphi^2 + p^2 d\lambda^2.$$

Now dS^2 determines the *metric*,

$$dS^2 = \sum_{i,j} g_{ij} dx^i dx^j = \mathbf{x}^\mathsf{T} \mathbf{H} \mathbf{x},$$

in which

$$g_{ij} = \begin{bmatrix} M^2 & 0 \\ 0 & p^2 \end{bmatrix} = \begin{bmatrix} E & F \\ F & G \end{bmatrix} = \mathbf{H}$$

and

$$\mathbf{x} = \begin{bmatrix} dx^1 \\ dx^2 \end{bmatrix} = \begin{bmatrix} d\varphi \\ d\lambda \end{bmatrix}.$$

In this, we have suitably defined two different notations, the index notation g_{ij}, dx^i and the matrix-vector notation $\mathbf{H}, \mathbf{x}$. The elements of the matrix are also the elements of the *first fundamental form of Gauss on the Earth's surface*, E, F, G. We see that $E = M^2, F = 0$ and $G = p^2$.

For the azimuth A again we obtain the following equation:

$$\tan A = \frac{p\, d\lambda}{M\, d\varphi},$$

from which, with the above,

$$dS \sin A = p\, d\lambda, \qquad dS \cos A = M\, d\varphi. \tag{15.2}$$

---

[6]Bernard H. Chovitz (1924–) is an American geodesist and cartographer.

15.3.2   In the map plane

If we project this little rectangle onto the map plane, we get the sides dx and dy, and, according to Pythagoras, its diagonal is

$$ds^2 = dx^2 + dy^2.$$

Now an element of distance in the *map plane* may be calculated as follows:

$$dx = \frac{\partial x}{\partial \varphi} d\varphi + \frac{\partial x}{\partial \lambda} d\lambda, \qquad dy = \frac{\partial y}{\partial \varphi} d\varphi + \frac{\partial y}{\partial \lambda} d\lambda,$$

i.e.,

$$ds^2 = \left(\frac{\partial x}{\partial \varphi} d\varphi + \frac{\partial x}{\partial \lambda} d\lambda \right)^2 + \left(\frac{\partial y}{\partial \varphi} d\varphi + \frac{\partial y}{\partial \lambda} d\lambda \right)^2 =$$
$$= \tilde{E} d\varphi^2 + 2\tilde{F} d\varphi d\lambda + \tilde{G} d\lambda^2, \tag{15.3}$$

in which

$$\tilde{E} = \left(\frac{\partial x}{\partial \varphi} \right)^2 + \left(\frac{\partial y}{\partial \varphi} \right)^2, \ \tilde{F} = \frac{\partial x}{\partial \varphi} \frac{\partial x}{\partial \lambda} + \frac{\partial y}{\partial \varphi} \frac{\partial y}{\partial \lambda}, \ \tilde{G} = \left(\frac{\partial x}{\partial \lambda} \right)^2 + \left(\frac{\partial y}{\partial \lambda} \right)^2.$$

$\tilde{E}, \tilde{F}$ and $\tilde{G}$ form the Gauss first fundamental form in the *map plane*.

If the element of distance in the map plane, ds^2, is interpreted as a *metric* of the Earth's surface, we obtain the metric tensor,

$$\widetilde{g_{ij}} = \begin{bmatrix} \tilde{E} & \tilde{F} \\ \tilde{F} & \tilde{G} \end{bmatrix} \overset{\text{def}}{=} \widetilde{\mathbf{H}}.$$

The corresponding metric is

$$ds^2 = \sum_{i,j} \widetilde{g_{ij}} dx^i dx^j = \mathbf{x}^\mathsf{T} \widetilde{\mathbf{H}} \mathbf{x},$$

in which $dx^1 = d\varphi$ and $dx^2 = d\lambda$, i.e., $dx^i = dx^j = \mathbf{x} = \begin{bmatrix} d\varphi & d\lambda \end{bmatrix}^\mathsf{T}$. Also here we see, as alternatives, the index notation and the matrix-vector notation, describing the same thing.

15.3.3   The scale

The *scale* is now the ratio

$$m = \frac{ds}{dS},$$

which obviously depends on the *direction* or azimuth A of the element of distance.

Write the *eigenvalue problem*:

$$m^2 = \frac{ds^2}{dS^2} = \frac{\sum_{i,j} \widetilde{g_{ij}} dx^i dx^j}{\sum_{i,j} g_{ij} dx^i dx^j} \implies \sum_{i,j} \widetilde{g_{ij}} dx^i dx^j - m^2 \sum_{i,j} g_{ij} dx^i dx^j = 0. \quad (15.4)$$

The same matric equation, if $\mathbf{x} = \begin{bmatrix} dx^1 & dx^2 \end{bmatrix}^\mathsf{T} = \begin{bmatrix} d\varphi & d\lambda \end{bmatrix}^\mathsf{T}$:

$$\mathbf{x}^\mathsf{T} \left(\widetilde{\mathbf{H}} - m^2 \mathbf{H} \right) \mathbf{x} = 0.$$

Equations 15.2 tell:

$$dS \sin A = p\, d\lambda, \qquad dS \cos A = M\, d\varphi.$$

Consider now all vectors of form

$$\mathbf{x} = \begin{bmatrix} d\varphi \\ d\lambda \end{bmatrix} = \begin{bmatrix} \cos A / M \\ \sin A / p \end{bmatrix}. \quad (15.5)$$

The lengths of these vectors on the surface of the Earth are:

$$dS^2 = M^2 d\varphi^2 + p^2 d\lambda^2 = M^2 \left(\frac{\cos A}{M} \right)^2 + p^2 \left(\frac{\sin A}{p} \right)^2 = \cos^2 A + \sin^2 A = 1.$$

So, on the Earth's surface these vectors form *a circle of unit radius*.

Substitute Equation 15.5 into Equation 15.4, using Equation 15.3:

$$\widetilde{E} \left(\frac{\cos A}{M} \right)^2 + \widetilde{G} \left(\frac{\sin A}{p} \right)^2 + 2\widetilde{F} \frac{\cos A}{M} \frac{\sin A}{p} -$$

$$- m^2 \left[M^2 \left(\frac{\cos A}{M} \right)^2 + p^2 \left(\frac{\sin A}{p} \right)^2 \right] = 0, \quad (15.6)$$

i.e., after cleaning up,

$$m^2 = \frac{\widetilde{E}}{M^2} \cos^2 A + \frac{\widetilde{G}}{p^2} \sin^2 A + 2\frac{\widetilde{F}}{Mp} \sin A \cos A =$$

$$= \tfrac{1}{2} \left(\frac{\widetilde{E}}{M^2} + \frac{\widetilde{G}}{p^2} \right) + \tfrac{1}{2} \left(\frac{\widetilde{E}}{M^2} - \frac{\widetilde{G}}{p^2} \right) \cos 2A + \frac{\widetilde{F}}{Mp} \sin 2A.$$

from this we obtain the *stationary values*:

$$0 = \frac{d}{dA} m^2 = \left(\frac{\widetilde{G}}{p^2} - \frac{\widetilde{E}}{M^2} \right) \sin 2A + 2\frac{\widetilde{F}}{Mp} \cos 2A,$$

i.e.,

$$\tan 2A = \frac{\dfrac{\tilde{E}}{M^2} - \dfrac{\tilde{G}}{p^2}}{\dfrac{2\tilde{F}}{Mp}} = \frac{\tilde{E}p^2 - \tilde{G}M^2}{2\tilde{F}Mp}.$$

This yields two maximum and two minimum values, which are all four at separations of 90° from each other[7].

These eigenvalues are obtained by writing Equation 15.6 as follows:

$$\begin{bmatrix} \cos A & \sin A \end{bmatrix} \begin{bmatrix} \dfrac{\tilde{E}}{M^2} - m^2 & \dfrac{\tilde{F}}{Mp} \\ \dfrac{\tilde{F}}{Mp} & \dfrac{\tilde{G}}{p^2} - m^2 \end{bmatrix} \begin{bmatrix} \cos A \\ \sin A \end{bmatrix} = 0.$$

This requires that the *determinant* of the matrix in the middle *vanishes*:

$$0 = \det\left(\mathbf{H}^{-1}\tilde{\mathbf{H}} - m^2\mathbf{I}\right) = \left(\frac{\tilde{E}}{M^2} - m^2\right)\left(\frac{\tilde{G}}{p^2} - m^2\right) - \frac{\tilde{F}^2}{M^2p^2} =$$

$$= m^4 + \left(-\frac{\tilde{E}}{M^2} - \frac{\tilde{G}}{p^2}\right)m^2 + \frac{1}{M^2p^2}\left(\tilde{E}\tilde{G} - \tilde{F}^2\right).$$

From this

$$m_{1,2}^2 = \frac{\left(\dfrac{\tilde{E}}{M^2} + \dfrac{\tilde{G}}{p^2}\right) \pm \sqrt{\left(\dfrac{\tilde{E}}{M^2} + \dfrac{\tilde{G}}{p^2}\right)^2 - 4\dfrac{1}{M^2p^2}\left(\tilde{E}\tilde{G} - \tilde{F}^2\right)}}{2}.$$

These two solutions m_1, m_2 are called the *principal scale factors*.

If the $\tilde{\mathbf{H}}$ matrix is a *diagonal matrix*:

$$\tilde{\mathbf{H}} = \begin{bmatrix} \tilde{E} & 0 \\ 0 & \tilde{G} \end{bmatrix},$$

the eigenvalues are obtained

$$\det\left(\mathbf{H}^{-1}\tilde{\mathbf{H}} - m^2\mathbf{I}\right) = \left(\frac{\tilde{E}}{M^2} - m^2\right)\left(\frac{\tilde{G}}{p^2} - m^2\right) = 0 \implies m_{1,2}^2 = \frac{\tilde{E}}{M^2}, \frac{\tilde{G}}{p^2}.$$

In this,

$$m_1 = \sqrt{\frac{\tilde{E}}{M^2}}, \qquad m_2 = \sqrt{\frac{\tilde{G}}{p^2}}$$

are the *meridional scale factor* and the scale factor in the direction of the parallel, or *transversal scale factor*. In intermediate directions, for azimuths A, the scale factor is then

$$m(A) = \sqrt{m_1^2 \cos^2 A + m_2^2 \sin^2 A}.$$

[7]If $F = 0$, the condition $\sin 2A = 0$ is obtained, which is fulfilled if $A = k \cdot 90°$, $k = 0, 1, 2, 3$.

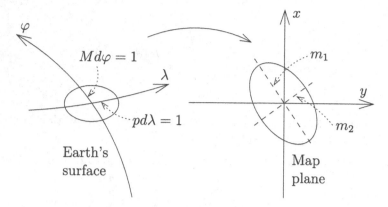

Figure 15.5: Tissot's indicatrix.

15.3.4 The Tissot indicatrix

The matrix ("scale tensor")

$$\mathbf{H}^{-1}\widetilde{\mathbf{H}} = \begin{bmatrix} \dfrac{\widetilde{E}}{M^2} & \dfrac{\widetilde{F}}{Mp} \\[2mm] \dfrac{\widetilde{F}}{Mp} & \dfrac{\widetilde{G}}{p^2} \end{bmatrix} = g^{ik}\widetilde{g}_{kj} = \widetilde{g}^i_j$$

is often visualized as an ellipse on the Earth's surface (or equivalently in the map plane). The eigenvalues of the matrix are m_1^2 and m_2^2. The semi-axes of the visualization ellipse are of lengths m_1 in one direction — often the meridional — and m_2 in the other, perpendicular direction — often the transversal direction. This ellipse is called the *Tissot*[8] *indicatrix* or distortion ellipse. See Figure 15.5 and Subsection 14.3.3 on page 214.

Of the many map projections used, we should mention especially those which are *conformal*, when the scale in a point is the same in all directions:

$$m_1 = m_2 = m.$$

In this case, the Tissot indicatrix is a *circle*. In conformal projections, small circles, squares and local angles and length ratios are mapped correctly. However, surface areas are distorted: an extreme example of this is Figure 1.6 on page 9.

Many conformal projections have the nice property that the projection equations for one of the co-ordinate directions can be derived, if the equations for the other co-ordinate direction are given. This concerns those projections for which either co-ordinate in the map plane is a function of only one geodetic co-ordinate, e.g., $y \to \lambda$ and $x \to \varphi$ in the case of Mercator.

Other map projections again have the property — a good one when mapping, e.g., the areal density of population or some other phenomenon — that surface

---

[8]Nicolas Auguste Tissot (1824–1897) was a French cartographer.

areas are mapped correctly, although the shapes of small circles or squares may be distorted. This condition, *equivalence*, is

$$\det\left(\mathbf{H}^{-1}\widetilde{\mathbf{H}}\right) = m_1^2 m_2^2 = \text{constant}.$$

15.4 Curvature of the Earth's surface and scale

Recall the equation derived in the previous Section 15.1 on page 233:

$$\mathcal{R} = -\left(\frac{\partial^2 g_{11}}{(\partial x^2)^2} + \frac{\partial^2 g_{22}}{(\partial x^1)^2}\right).$$

Because, according to Equation 14.11 on page 222, $\mathcal{R} = 2K$, one obtains

$$K = -\frac{1}{2}\left(\frac{\partial^2 g_{11}}{(\partial x^2)^2} + \frac{\partial^2 g_{22}}{(\partial x^1)^2}\right).$$

Now take the rectangular co-ordinate frame (x, y) *in the map plane* and move its co-ordinate lines back onto the curved surface of the Earth, forming there a curvilinear co-ordinate frame (ξ, η). At the origin of the map plane, or at the central meridian, the metric g_{ij} of this co-ordinate frame is, at least in the case of ordinary map projections, *quasi-Cartesian*, i.e., the metric is the unit matrix and stationary at the origin. In this case, the theory of the previous chapter (Section 15.2 on page 232) applies.

In case of a *conformal projection*, the form of this metric is

$$g_{ij} = m^{-2}\begin{bmatrix} 1 & 0 \\ 0 & 1 \end{bmatrix},$$

in which m is the scale of the map projection, which thus depends on the place x^i.

Compute

$$K = -\frac{1}{2}\left(\frac{\partial^2 g_{\xi\xi}}{\partial \eta^2} + \frac{\partial^2 g_{\eta\eta}}{\partial \xi^2}\right) = -\frac{1}{2}\Delta\left(m^{-2}\right).$$

Now

$$\Delta\left(m^{-2}\right) = \left(\frac{\partial^2}{\partial \xi^2} + \frac{\partial^2}{\partial \eta^2}\right)m^{-2}(\xi, \eta) =$$

$$= -2\frac{\partial}{\partial \xi}\left(m^{-3}\frac{\partial m}{\partial \xi}\right) - 2\frac{\partial}{\partial \eta}\left(m^{-3}\frac{\partial m}{\partial \eta}\right) =$$

$$= -2\left(\frac{\partial m^{-3}}{\partial \xi}\cdot\frac{\partial m}{\partial \xi} + m^{-3}\frac{\partial^2 m}{\partial \xi^2}\right) - 2\left(\frac{\partial m^{-3}}{\partial \eta}\cdot\frac{\partial m}{\partial \eta} + m^{-3}\frac{\partial^2 m}{\partial \eta^2}\right)$$

$$\approx -2\Delta m,$$

considering the stationarity of m (and thus of m^{-3}), and $m \approx 1$. So

$$K = \Delta m.$$

So, there is a simple relationship between the Gaussian curvature and the second derivative of place of the scale (more precisely, the Laplace operator Δ). As a result, there is also a connection between the second derivatives of the scale in the North-South and East-West directions. If, e.g.,

$$\frac{\partial^2 m}{\partial \xi^2} = 0 \implies \frac{\partial^2 m}{\partial \eta^2} = K$$

etc. The below table gives some examples. Recall that the area considered is always the surroundings of the central point, meridian, standard parallel or equator, where $m \approx 1$ and stationary!

Projection	$\partial^2 m / \partial x^2$	$\partial^2 m / \partial y^2$
Mercator	K	0
Lambert conical	K	0
Oblique stereographic	$K/2$	$K/2$
Gauss–Krüger, UTM	0	K

In the table, we have again used instead of (ξ, η), (x, y), i.e., made the substitutions

$$\frac{\partial^2 m}{\partial \xi^2} \to \frac{\partial^2 m}{\partial x^2}, \qquad \frac{\partial^2 m}{\partial \eta^2} \to \frac{\partial^2 m}{\partial y^2},$$

which is allowed based on quasi-Cartesianness.

The scale of the Mercator projection is constant in the direction of the y co-ordinate, i.e., from map West to map East. The scale of transversal cylindrical projections again is constant in the direction of the x co-ordinate or central meridian. The stereographic oblique projection again is symmetric and the scale behaves in the same way in all directions from the centre point of the map.

For this reason, the second derivative of the scale may be used for a classification of map projections. However, as said already, one should also require the first derivatives of the scale, $\dfrac{\partial m}{\partial x}$ and $\dfrac{\partial m}{\partial y}$, to vanish in the centre of the area mapped: e.g., the first derivative in the x direction of the classical Mercator projection vanishes only on the equator, whereas in the case of the Lambert projection it vanishes for the reference latitude φ_0. Thus, Mercator is not suitable as a map projection for a high-latitude country, whereas Lambert, for an appropriately chosen reference latitude, is suitable.

As examples, Lambert is best suited for a country extending in the East-West direction (Estonia), whereas Gauss–Krüger is best for a country extending in the North-South direction (like Finland). The conformal azimuthal projection known as the oblique stereographic projection is suitable for a country in the shape of a square (the Netherlands).

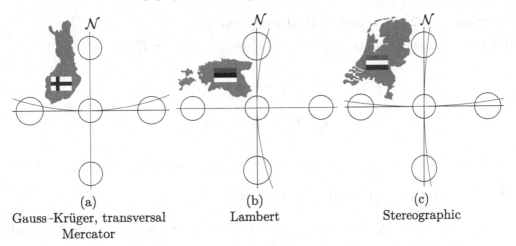

(a)
Gauss–Krüger, transversal
Mercator

(b)
Lambert

(c)
Stereographic

Figure 15.6: Classification of map projections into vertical, horizontal and "square" projections.

Exercises

Exercise 15 – 1: The Mercator projection

The projection formulas are (sphere):

$$x = R\lambda,$$
$$y = R \ln \tan\left(\frac{\pi}{4} + \frac{\varphi}{2}\right).$$

a. Calculate the Gauss first fundamental form for this map projection,

$$\widetilde{E} = \left(\frac{\partial x}{\partial \varphi}\right)^2 + \left(\frac{\partial y}{\partial \varphi}\right)^2,$$
$$\widetilde{F} = \frac{\partial x}{\partial \varphi}\frac{\partial x}{\partial \lambda} + \frac{\partial y}{\partial \varphi}\frac{\partial y}{\partial \lambda},$$
$$\widetilde{G} = \left(\frac{\partial x}{\partial \lambda}\right)^2 + \left(\frac{\partial y}{\partial \lambda}\right)^2$$

b. Calculate the principal scale factors. Is this a conformal projection?

c. In what way does the projection

$$x = R\lambda,$$
$$y = -R \ln \tan\left(\frac{\pi}{4} - \frac{\varphi}{2}\right)$$

differ from the given projection?

Exercise 15 – 2: An azimuthal projection

Given the following map projection (sphere, i.e., $M = R, p = R\cos\varphi$):

$$x = R\left(\frac{\pi}{2} - \varphi\right)\cos\lambda,$$

$$y = R\left(\frac{\pi}{2} - \varphi\right)\sin\lambda.$$

a. Calculate the Gauss fundamental form in the map plane, $\tilde{E}, \tilde{F}, \tilde{G}$.

b. Calculate the principal scale factors m_1, m_2.

c. Is this a conformal projection? Is it equivalent (area true)? Is it equidistant?

Exercise 15 – 3: Yet another projection

Given still the following map projection (also on the sphere):

$$x = R\varphi,$$

$$y = R\lambda\cos\varphi.$$

a. Calculate the Gauss fundamental form in the map plane, $\tilde{E}, \tilde{F}, \tilde{G}$.

b. Calculate the matrix

$$\mathbf{M} \overset{\text{def}}{=} \widetilde{\mathbf{H}}^{-1}\mathbf{H} = \begin{bmatrix} \frac{\tilde{E}}{M^2} & \frac{\tilde{F}}{Mp} \\ \frac{\tilde{F}}{Mp} & \frac{G}{p^2} \end{bmatrix},$$

("scale tensor"), in the case of a sphere, i.e., $M = R, p = R\cos\varphi$.

c. Calculate the *determinant* of the matrix $\mathbf{M}$.

d. Remember that $\det(\mathbf{M}) = m_1^2 m_2^2$. Is this an equivalent projection?

Exercise 15 – 4: Spherical triangle, spherical excess

The corner points of a spherical triangle are

$$\varphi_1 = 60°, \quad \lambda_1 = 25°,$$
$$\varphi_2 = 65°, \quad \lambda_2 = 27°,$$
$$\varphi_3 = 62°, \quad \lambda_3 = 28°.$$

a. Calculate the vectors $\mathbf{x}_1, \mathbf{x}_2, \mathbf{x}_3$, from these $\mathbf{y}_1, \mathbf{y}_2, \mathbf{y}_3$, and in this way, the spherical excess.

b. Calculate the lengths of the sides of triangle (123).

Exercise 15 – 5: The Chovitz curvature condition

What is the more general form of the Chovitz curvature condition (Equation 15.1 on page 233) in the case where we do *not* assume the co-ordinate curves to be orthogonal?

Appendix A

The stereographic projection maps circles into circles

See Figure A.1. In the proof, we exploit the fact that both the surface of the sphere and the map plane are inside $\mathbb{R}^3$. According to Figure 5.3 on page 60, the inversion through a circle in two-dimensional space $\mathbb{R}^2$ maps general circles (i.e., either circles or straight lines) into general circles. The formula for the inversion is at its simplest (unit circle, inversion centre at the origin)

$$f(\mathbf{x}) = \mathbf{x} \, \|\mathbf{x}\|^{-2} \, ,$$

which will work also in higher dimensions. For example, in $\mathbb{R}^3$ this describes the inversion through the unit sphere. By rotating Figure 5.3 on page 60 around its symmetry axis c we see immediately that this inversion *maps spheres into spheres* also in three dimensions. "Sphere" here stands for "general sphere"; either a sphere or a plane.

In Figure A.1 we see that the map plane K is the image of the spherical surface M through the inversion sphere. Let P_1 be an arbitrary sphere which

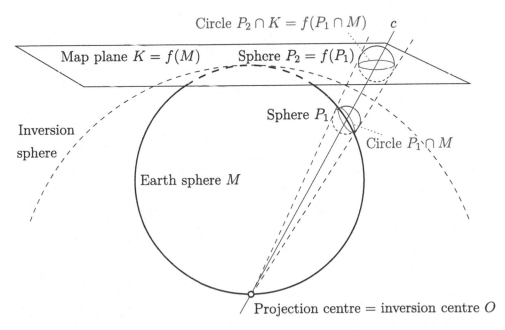

Figure A.1: The stereographic projection as a restriction of the inversion in $\mathbb{R}^3$ through a sphere.

intersects M in a circle[1] $P_1 \cap M$. If now $P_2 = f(P_1)$ is the image of P_1 mapped through the same inversion through the sphere, it follows directly that this sphere intersects the map plane *in the circle* $P_2 \cap K$, which is the same as the image of the original circle $P_1 \cap M$ mapped by this inversion through the sphere!

So,

An inversion through the sphere in $\mathbb{R}^3$, restricted to a spherical surface and a map plane, is a stereographic projection.

---

[1]The intersection of two general spheres is always a general circle.

Appendix B

Isometric latitude on the ellipsoid

We follow the presentation from the book (Grossman, 1964).
The starting equation is

$$\psi(\varphi) = \int_0^\varphi \frac{M(\varphi')}{p(\varphi')} d\varphi'.$$

The differential-equation version of this is

$$d\psi = \frac{M}{p} d\varphi = \frac{1 - e^2}{(1 - e^2 \sin^2 \varphi) \cos \varphi} d\varphi.$$

The integrand is split up into partial fractions:

$$\frac{1 - e^2}{(1 - e^2 \sin^2 \varphi) \cos \varphi} = \frac{1 - e^2 \sin^2 \varphi - e^2 \cos^2 \varphi}{(1 - e^2 \sin^2 \varphi) \cos \varphi} =$$

$$= \frac{1}{\cos \varphi} - \frac{e^2 \cos \varphi}{1 - e^2 \sin^2 \varphi} =$$

$$= \frac{1}{\cos \varphi} - \frac{e^2 \cos \varphi (1 - e \sin \varphi) + e^2 \cos \varphi (1 + e \sin \varphi)}{2 (1 + e \sin \varphi) (1 - e \sin \varphi)} =$$

$$= \frac{1}{\cos \varphi} + \frac{e}{2} \left(-\frac{e \cos \varphi}{1 + e \sin \varphi} - \frac{e \cos \varphi}{1 - e \sin \varphi} \right).$$

The integral of the first term is

$$\psi = \int_0^\varphi \frac{1}{\cos \varphi'} d\varphi' = \ln \tan \left(\frac{\pi}{4} + \frac{\varphi}{2} \right).$$

Proof using the chain rule:

$$\frac{d\psi}{d\varphi} = \frac{d \ln \tan \left(\frac{\pi}{4} + \frac{\varphi}{2} \right)}{d \tan \left(\frac{\pi}{4} + \frac{\varphi}{2} \right)} \cdot \frac{d \tan \left(\frac{\pi}{4} + \frac{\varphi}{2} \right)}{d \left(\frac{\pi}{4} + \frac{\varphi}{2} \right)} \cdot \frac{d \left(\frac{\pi}{4} + \frac{\varphi}{2} \right)}{d\varphi} =$$

$$= \frac{1}{\tan \left(\frac{\pi}{4} + \frac{\varphi}{2} \right)} \cdot \frac{1}{\cos^2 \left(\frac{\pi}{4} + \frac{\varphi}{2} \right)} \cdot \frac{1}{2} =$$

$$= \frac{1}{2 \sin \left(\frac{\pi}{4} + \frac{\varphi}{2} \right) \cos \left(\frac{\pi}{4} + \frac{\varphi}{2} \right)} = \frac{1}{\sin \left(\frac{\pi}{2} + \varphi \right)} = \frac{1}{\cos \varphi}.$$

This is *the whole solution* in the case $e = 0$ (spherical solution).

In the case of the ellipsoid, the second integral

$$\int \left(-\frac{e\cos\varphi}{1 + e\sin\varphi}\right) d\varphi = -\int \frac{f'(\varphi)}{f(\varphi)} d\varphi = -\ln f(\varphi) = -\ln(1 + e\sin\varphi),$$

in which we have used the notation $f(\varphi) \overset{\text{def}}{=} 1 + e\sin\varphi$. Similarly

$$\int \left(-\frac{e\cos\varphi}{1 - e\sin\varphi}\right) d\varphi = \ln(1 - e\sin\varphi),$$

and the end result is

$$\psi = \ln\tan\left(\frac{\pi}{4} + \frac{\varphi}{2}\right) + \frac{e}{2}\left(\ln(1 - e\sin\varphi) - \ln(1 + e\sin\varphi)\right) =$$
$$= \ln\left(\tan\left(\frac{\pi}{4} + \frac{\varphi}{2}\right)\left(\frac{1 - e\sin\varphi}{1 + e\sin\varphi}\right)^{e/2}\right).$$

Appendix C

Useful relations between the principal radii of curvature

Given the equations for the principal radii of curvature of the ellipsoid of revolution

$$N(\varphi) = a \left(1 - e^2 \sin^2 \varphi \right)^{-1/2},$$
$$M(\varphi) = a \left(1 - e^2 \right) \left(1 - e^2 \sin^2 \varphi \right)^{-3/2}, \tag{C.1}$$

we may derive by differentiation:

$$\frac{d}{d\varphi} \left(N(\varphi) \cos \varphi \right) = -M(\varphi) \sin \varphi,$$
$$\frac{d}{d\varphi} \left(N(\varphi) \sin \varphi \right) = +\frac{M(\varphi)}{1 - e^2} \cos \varphi.$$

Furthermore

$$\frac{d}{d\varphi} M^2(\varphi) = a^2 \left(1 - e^2 \right)^2 \frac{d}{d\varphi} \left(1 - e^2 \sin^2 \varphi \right)^{-3} =$$
$$= 3 \frac{e^2 M^2 N^2}{a^2} \sin 2\varphi,$$
$$\frac{d}{d\varphi} \left(N^2(\varphi) \cos^2 \varphi \right) = 2N \cos \varphi \frac{d}{d\varphi} \left(N(\varphi) \cos \varphi \right) =$$
$$= -MN \sin 2\varphi.$$

Appendix D

Christoffel's symbols from the metric

We start from the definition of the metric in Chapter 13.2 on page 184:

$$g_{ij} = \left\langle \frac{\partial \mathbf{x}}{\partial x^i} \cdot \frac{\partial \mathbf{x}}{\partial x^j} \right\rangle = \begin{bmatrix} \left\langle \frac{\partial \mathbf{x}}{\partial x^1} \cdot \frac{\partial \mathbf{x}}{\partial x^1} \right\rangle & \left\langle \frac{\partial \mathbf{x}}{\partial x^1} \cdot \frac{\partial \mathbf{x}}{\partial x^2} \right\rangle \\ \left\langle \frac{\partial \mathbf{x}}{\partial x^2} \cdot \frac{\partial \mathbf{x}}{\partial x^1} \right\rangle & \left\langle \frac{\partial \mathbf{x}}{\partial x^2} \cdot \frac{\partial \mathbf{x}}{\partial x^2} \right\rangle \end{bmatrix},$$

in which $x^i = (x^1, x^2)$ is a parametrization ("co-ordinate frame") of a curved surface (or, in the general case, some lower-dimensional curved entity) embedded in the space of the vector $\mathbf{x}$. Let us differentiate:

$$\frac{\partial g_{jk}}{\partial x^i} = \frac{\partial}{\partial x^i} \left\langle \frac{\partial \mathbf{x}}{\partial x^j} \cdot \frac{\partial \mathbf{x}}{\partial x^k} \right\rangle =$$

$$= \left\langle \frac{\partial^2 \mathbf{x}}{\partial x^i \partial x^j} \cdot \frac{\partial \mathbf{x}}{\partial x^k} \right\rangle + \left\langle \frac{\partial^2 \mathbf{x}}{\partial x^i \partial x^k} \cdot \frac{\partial \mathbf{x}}{\partial x^j} \right\rangle. \tag{D.1}$$

Interchange the three indices:

$$\frac{\partial g_{ki}}{\partial x^j} = \left\langle \frac{\partial^2 \mathbf{x}}{\partial x^j \partial x^k} \cdot \frac{\partial \mathbf{x}}{\partial x^i} \right\rangle + \left\langle \frac{\partial^2 \mathbf{x}}{\partial x^j \partial x^i} \cdot \frac{\partial \mathbf{x}}{\partial x^k} \right\rangle \tag{D.2}$$

$$\frac{\partial g_{ij}}{\partial x^k} = \left\langle \frac{\partial^2 \mathbf{x}}{\partial x^k \partial x^i} \cdot \frac{\partial \mathbf{x}}{\partial x^j} \right\rangle + \left\langle \frac{\partial^2 \mathbf{x}}{\partial x^k \partial x^j} \cdot \frac{\partial \mathbf{x}}{\partial x^i} \right\rangle \tag{D.3}$$

Calculate Equation D.1 plus Equation D.2 minus Equation D.3:

$$\frac{\partial g_{jk}}{\partial x^i} + \frac{\partial g_{ki}}{\partial x^j} - \frac{\partial g_{ij}}{\partial x^k} = 2 \left\langle \frac{\partial^2 \mathbf{x}}{\partial x^i \partial x^j} \cdot \frac{\partial \mathbf{x}}{\partial x^k} \right\rangle. \tag{D.4}$$

Express the second derivatives of $\mathbf{x}$ on the local basis

$$\left\{ \frac{\partial \mathbf{x}}{\partial x^1}, \frac{\partial \mathbf{x}}{\partial x^2}, \mathbf{n} \right\}$$

as was done in Equation 13.8 on page 193, albeit in a slightly differing notation:

$$\frac{\partial^2 \mathbf{x}}{\partial x^i \partial x^j} \overset{\text{def}}{=} \Gamma^\ell_{ij} \frac{\partial \mathbf{x}}{\partial x^\ell} + \beta_{ij} \mathbf{n}, \tag{D.5}$$

where implicitly are defined the Γ symbols. Substitution into Equation D.4 yields

$$\Gamma_{ij}^{\ell} \left\langle \frac{\partial \mathbf{x}}{\partial x^{\ell}} \cdot \frac{\partial \mathbf{x}}{\partial x^{k}} \right\rangle + \beta_{ij} \left\langle \mathbf{n} \cdot \frac{\partial \mathbf{x}}{\partial x^{k}} \right\rangle = \frac{1}{2} \left(\frac{\partial}{\partial x^{i}} g_{jk} + \frac{\partial}{\partial x^{j}} g_{ki} - \frac{\partial}{\partial x^{k}} g_{ij} \right).$$

Here we identify

$$\left\langle \frac{\partial \mathbf{x}}{\partial x^{\ell}} \cdot \frac{\partial \mathbf{x}}{\partial x^{k}} \right\rangle = g_{\ell k} \quad \text{and} \quad \left\langle \mathbf{n} \cdot \frac{\partial \mathbf{x}}{\partial x^{k}} \right\rangle = 0,$$

i.e.,

$$\Gamma_{ij}^{\ell} \, g_{\ell k} = \frac{1}{2} \left(\frac{\partial g_{jk}}{\partial x^{i}} + \frac{\partial g_{ki}}{\partial x^{j}} - \frac{\partial g_{ij}}{\partial x^{k}} \right)$$

$$\implies \Gamma_{ij}^{\ell} = \frac{1}{2} g^{\ell k} \left(\frac{\partial g_{jk}}{\partial x^{i}} + \frac{\partial g_{ki}}{\partial x^{j}} - \frac{\partial g_{ij}}{\partial x^{k}} \right),$$

Equation 14.6 on page 216. Here, $g^{ij} g_{jk} = \delta_{k}^{i}$, i.e., g^{ij} is the inverse matrix of g_{ij}.

Appendix E

The Riemann tensor from the Christoffel symbols

The Riemann tensor formula is derived with the aid of parallel transport of a vector around a closed co-ordinate square.

If the vector in space $\mathbf{v}$ is transported in a parallel fashion within the surface S parametrized by x^i, we have

$$\frac{\partial \mathbf{v}}{\partial x^i} = 0.$$

Write $\mathbf{v}$ on a basis of tangent vectors:

$$\mathbf{v} = v^i \frac{\partial \mathbf{x}}{\partial x^i}.$$

Then

$$0 = \frac{\partial \mathbf{v}}{\partial x^j} = \frac{\partial v^i}{\partial x^j} \frac{\partial \mathbf{x}}{\partial x^i} + v^i \frac{\partial^2 \mathbf{x}}{\partial x^i \partial x^j} =$$

$$= \frac{\partial v^i}{\partial x^j} \frac{\partial \mathbf{x}}{\partial x^i} + \Gamma^i_{jk} v^k \frac{\partial \mathbf{x}}{\partial x^i} + \beta_{ij} \mathbf{n},$$

using Equation D.5 on page 251. In this, the $\mathbf{v}$ derivative consists of two parts: an "interior" part,

$$\frac{\partial v^i}{\partial x^j} \frac{\partial \mathbf{x}}{\partial x^i} + \Gamma^i_{jk} v^k \frac{\partial \mathbf{x}}{\partial x^i},$$

inside the surface, and an "exterior" part, $\beta_{ij}\mathbf{n}$, which is perpendicular to the surface. When the surface S is given, we may just zero the interior part, i.e.,

$$\frac{\partial v^i}{\partial x^j} + \Gamma^i_{jk} v^k = 0 \tag{E.1}$$

describes the parallel transport of the vector v^i within the surface.

Let us now describe a small quadrilateral $ABCD$, with side lengths Δx^k and Δx^ℓ (see Figure 14.6 on page 219), along co-ordinate curves. The sides AB and CD are on opposite sides, the running co-ordinate being x^k. Similarly BC and AD are opposite, and their running co-ordinate is x^ℓ.

Compute the change in v^i over the transport distance AB:

$$\Delta_{AB} v^i = \frac{\partial v^i}{\partial x^k} \Delta x^k = -\Gamma^i_{km} v^m \Delta x^k.$$

Similarly

$$\Delta_{CD} v^i = +\Gamma^i_{km} v^m \Delta x^k.$$

For side BC we obtain

$$\Delta_{BC} v^i = \frac{\partial v^i}{\partial x^\ell} \Delta x^\ell = -\Gamma^i_{\ell m} v^m \Delta x^\ell$$

and

$$\Delta_{DA} v^i = +\Gamma^i_{\ell m} v^m \Delta x^\ell.$$

Add these four terms together:

$$\Delta_{ABCD} v^i = \left(\left(\Gamma^i_{km} v^m \right)_{CD} - \left(\Gamma^i_{km} v^m \right)_{AB} \right) \Delta x^k -$$

$$- \left(\left(\Gamma^i_{\ell m} v^m \right)_{DA} - \left(\Gamma^i_{\ell m} v^m \right)_{BC} \right) \Delta x^\ell =$$

$$\approx \left(\frac{\partial}{\partial x^\ell} \left(\Gamma^i_{km} v^m \right) \Delta x^\ell \right) \Delta x^k - \left(\frac{\partial}{\partial x^k} \left(\Gamma^i_{\ell m} v^m \right) \Delta x^k \right) \Delta x^\ell =$$

$$= \left(\left(\frac{\partial \Gamma^i_{km}}{\partial x^\ell} - \frac{\partial \Gamma^i_{\ell m}}{\partial x^k} \right) v^m + \Gamma^i_{km} \frac{\partial v^m}{\partial x^\ell} - \Gamma^i_{\ell m} \frac{\partial v^m}{\partial x^k} \right) \Delta x^\ell \Delta x^k.$$

Equation E.1 yields

$$\frac{\partial v^m}{\partial x^\ell} = -\Gamma^m_{\ell h} v^h, \ \frac{\partial v^m}{\partial x^k} = -\Gamma^m_{kh} v^h.$$

Substitute:

$$\Delta_{ABCD} v^i = \left(\left(\frac{\partial \Gamma^i_{km}}{\partial x^\ell} - \frac{\partial \Gamma^i_{\ell m}}{\partial x^k} \right) v^m + \left(\Gamma^i_{\ell m} \Gamma^m_{kh} - \Gamma^i_{km} \Gamma^m_{\ell h} \right) v^h \right) \Delta x^\ell \Delta x^k =$$

$$= \left(\frac{\partial \Gamma^i_{kj}}{\partial x^\ell} - \frac{\partial \Gamma^i_{\ell j}}{\partial x^k} + \Gamma^i_{\ell m} \Gamma^m_{kj} - \Gamma^i_{km} \Gamma^m_{\ell j} \right) v^j \Delta x^\ell \Delta x^k,$$

in which we have changed the names of the indices, $m \to j$ (in the former two terms) and $h \to j$ (in the latter two terms).

From this we see the *Riemann curvature tensor*:

$$R^i_{j\ell k} = \frac{\partial \Gamma^i_{kj}}{\partial x^\ell} - \frac{\partial \Gamma^i_{\ell j}}{\partial x^k} + \Gamma^i_{\ell m} \Gamma^m_{kj} - \Gamma^i_{km} \Gamma^m_{\ell j},$$

apart from the names of the indices, and changes of type $\Gamma^i_{jk} = \Gamma^i_{kj}$, this tensor is precisely the one already given in Equation 14.9 on page 220.

Appendix F

Conformal mappings in spaces of any dimension

Let $D \subset \mathbb{R}^n$ be a domain, $n \geq 2$ and given the mapping

$$\mathbf{y} = \mathbf{F}(\mathbf{x}), \qquad \mathbf{x} \in D, \mathbf{y} \in \mathbb{R}^n,$$

the partial derivatives of all component functions of which are defined on the domain D. Then, the mapping of small difference vectors in a neighbourhood of point $\mathbf{x} \in D$ is described by the *matrix of Jacobi*[1,2] of the function $\mathbf{F}$, $\mathbf{F}'(\mathbf{x})$:

$$\mathbf{y_2} - \mathbf{y_1} \approx \mathbf{F}'(\mathbf{x})(\mathbf{x_2} - \mathbf{x_1}), \quad \mathbf{F}'(\mathbf{x}) = \begin{bmatrix} \dfrac{\partial y^1}{\partial x^1} & \cdots & \dfrac{\partial y^1}{\partial x^n} \\ \vdots & \ddots & \vdots \\ \dfrac{\partial y^n}{\partial x^1} & \cdots & \dfrac{\partial y^n}{\partial x^n} \end{bmatrix},$$

and infinitesimally

$$d\mathbf{y} = \mathbf{F}'(\mathbf{x})d\mathbf{x}, \tag{F.1}$$

in which

$$[\mathbf{F}']^i_j = \frac{\partial y^i}{\partial x^j}.$$

Let us study what happens to the length of distance $d\mathbf{x}$. Let the *norm* of the matrix $\mathbf{F}'$ be $\|\mathbf{F}'\|$, i.e., the absolutely smallest number $c \in \mathbb{R}$ for which the inequality

$$\|d\mathbf{y}\| \leq c \cdot \|d\mathbf{x}\| \tag{F.2}$$

holds. The number c is thus the largest "stretch" or norm of the directional derivative, which the mapping $\mathbf{F}$ produces locally, close to point $\mathbf{x}$.

A mapping is called *conformal*, if the inequality F.2 is the identity

$$\|d\mathbf{y}\| = c \cdot \|d\mathbf{x}\| \tag{F.3}$$

independently of the direction of the distance $d\mathbf{x}$, assuming $\det \mathbf{F}'(\mathbf{x}) > 0$. This corresponds to the intuitive concept of conformality. Note that, if $n = 2$ and $\mathbf{F}$ is

---

[1]Carl Gustav Jacob Jacobi (1804–1851) was a German mathematician, the discoverer of, among other things, the matrix and determinant of Jacobi.

[2]For the matrix of Jacobi, often the notation $\mathbf{F}'(\mathbf{x}) = \mathbf{J}_F(\mathbf{x})$ is used.

conformal, then the identity F.3 follows immediately from the Cauchy–Riemann conditions.

We can readily show that a conformal mapping preserves angles; e.g., right angles are mapped into right angles. Let the angle between two vectors dx_1 and dx_2 be $\alpha = 90°$ and $\|dx_1\| = \|dx_2\| = s$. Then

$$\|dx_1 + dx_2\| = \|dx_1 - dx_2\| = s\sqrt{2}.$$

If the angle α' between the image vectors $dy_1 = F'(x)dx_1$ and $dy_2 = F'(x)dx_2$ is however *not* a right angle, we obtain

$$\|dy_1 + dy_2\|^2 = \|dy_1\|^2 + \|dy_2\|^2 + 2\|dy_1\|\|dy_2\|\cos\alpha',$$

$$\|dy_1 - dy_2\|^2 = \|dy_1\|^2 + \|dy_2\|^2 - 2\|dy_1\|\|dy_2\|\cos\alpha',$$

which are *not* of equal length if $\alpha' \neq 90°$, violating the above conformality condition.

If we fix x, then the Jacobi matrix $F'(x)$ is an ordinary square matrix of size $n \times n$. Such a matrix may be written out as a singular-value decomposition (SVD), $F' = R\Lambda S$, in which the matrices R and S may always be chosen to be orthogonal (see Section 9.3 on page 112) and Λ is a diagonal matrix. Then we obtain $\det R = \det S = 1$, i.e., $\det F' = \det \Lambda$, and the latter is simply the product of the diagonal elements. We can now write Equation F.1 as follows:

$$dy' = \Lambda(x)dx',$$

in which $dy = Rdy'$ and $dx = S^T dx' = S^{-1}dx'$.

In the conformal case, the diagonal elements of Λ must all be the same, say c, in which case the norm of the matrix F' will also be c — as orthogonal matrices do not cause any deformation. It follows that

$$\|F'\|^n = \|\Lambda\|^n = \det \Lambda = \det F',$$

i.e., we obtain the conformality condition 5.4 on page 62.

What is left is the matter of *orientation*, which has to do with the algebraic sign of the determinant of Jacobi. If it is negative, we have a mirroring or orientation changing mapping, which in $\mathbb{R}^2$ maps a clockwise circular motion onto a counterclockwise one, in $\mathbb{R}^3$, a right turning corkscrew or a right hand into a left turning one or left hand, etc. In other words, oriented (hyper-) volumes change their orientation under the mapping.

In cartography, the issue is without significance, although globes of the celestial sphere have been manufactured that show the constellations — unlike in a planetarium — how they would look "from the outside," like a God's eye view[3]. This is an example of an orientation-changing mapping.

---

[3]The *Burgerzaal* in the Amsterdam Royal Palace has a beautiful stereographic star chart embedded in the marble floor, which is of this type.

References

Lars V. Ahlfors. *Complex Analysis*. MacGraw-Hill, New York, 1966.

Kenneth Auchincloss. Smithsonian Astronomers Keep Hectic Pace. *The Harvard Crimson*, November 9, 1957. URL http://www.thecrimson.com/article/1957/11/9/smithsonian-astronomers-keep-hectic-pace-pwere/. Student newsletter. Accessed March 23, 2019.

Aatish Bhatia. How a 19th century math genius taught us the best way to hold a pizza slice. *Wired*, September 5, 2014. URL https://www.wired.com/2014/09/curvature-and-strength-empzeal/. Accessed March 23, 2019.

Guy Bomford. Report of Study Group No. 10. The Geoid in Europe and Connected Countries. In *Traveaux de l'Association Internationale de Géodésie*, volume 22, 1963.

Claude Boucher and Zuhair Altamimi. Specifications for Reference Frame Fixing in the Analysis of a EUREF GPS Campaign. In *VI Meeting of the TWG in Bern, March 9 and 10, Veröffentlichungen der Bayerischen Kommission für die Internationale Erdmessung*, number 56, pages 265–267, München, 1995.

Claude Boucher and Zuheir Altamimi. *Specifications for reference frame fixing in the analysis of a EUREF GPS campaign*, 2008. URL http://etrs89.ensg.ign.fr/memo-V7.pdf. Version 7, 24-10-2008. Accessed April 3, 2019.

Claude Boucher, Zuheir Altamimi, and Patrick Sillard. Results and Analysis of the ITRF96 (IERS Technical Note No. 24). Technical report, Central Bureau of IERS – Observatoire de Paris, Paris, 1998. URL http://www.iers.org/sid_A8C345E590AABEF7D53F7B7339DD7D5C/IERS/EN/Publications/TechnicalNotes/tn24.html.

Sean Carroll. *Lecture Notes on General Relativity*. arXiv, 1997. URL https://arxiv.org/pdf/gr-qc/9712019v1.pdf.

Bernard H. Chovitz. Classification of map projections in terms of the metric tensor to the second order. *Bollettino di Geodesia e Scienze Affini*, 11, 1952.

Nicholas Crane. *Mercator, the man who mapped the planet*. Weidenfeld & Nicolson, London, 2002. ISBN 0 297 64665 6.

Tobin A. Driscoll. The Schwarz-Christoffel Toolbox for MATLAB, maintained. URL http://www.math.udel.edu/~driscoll/SC/. Accessed March 23, 2019.

Tobin A. Driscoll and Lloyd N. Trefethen. *Schwarz-Christoffel Mapping*. Cambridge University Press, Cambridge, 2002.

Graham Farmelo. *The Strangest Man*. Basic Books, reprint edition, 2011. ISBN 978-0-4650-2210-6.

James Gall. Uses of Cylindrical Projections for Geographical, Astronomical, and Scientific Purposes. *Scottish Geographical Magazine*, 1(4):119–123, 1885. URL http://www.heliheyn.de/Maps/JamesGall/Article.htm.

Carl Friedrich Gauss. Teoria Residuorum Biquadraticorum, Commentatio Secunda. In *Werke, Band II*. Köningliche Gesellschaft der Wissenschaften, Göttingen, second edition, 1876. URL http://www.wilbourhall.org/pdfs/Carl_Friedrich_Gauss_Werke___2.pdf. PDF page 92, page 110.

Frederick W. Gehring. Rings and quasiconformal mappings in space. *Transactions of the American Mathematical Society*, 103:353–393, 1962. ISSN 0002-9947. URL http://www.ams.org/tran/1962-103-03/S0002-9947-1962-0139735-8/S0002-9947-1962-0139735-8.pdf.

Richard S. Gross. The excitation of the Chandler wobble. *Geophysical Research Letters*, 27:2329–2332, 2000.

Walter Grossman. *Geodätische Rechnungen und Abbildungen*. Verlag Konrad Wittwer, Stuttgart, 1964.

N. D. Haasbroek. Gemma Frisius, Tycho Brahe and Snellius and their triangulations. Technical report, Netherlands Geodetic Commission, Delft, 1968. URL https://ncgeo.nl/index.php/nl/publicaties/groene-serie/item/2332-gs-14-n-d-haasbroek-gemma-frisius-tycho-brahe-and-snellius-and-their-triangulations. Accessed April 3, 2019.

Harri Hakula, Tri Quach, and Antti Rasila. Conjugate function method for numerical conformal mappings. *Journal of Computational and Applied Mathematics*, pages 340–353, 2013. doi: 10.1016/j.cam.2012.06.003.

Philip Hartman. On isometries and on a theorem of Liouville. *Mathematische Zeitschrift*, 69:202–210, 1958. ISSN 0025-5874.

Heavens Above, maintained, Chris Peat. URL https://www.heavens-above.com/. Accessed March 23, 2019.

Markku Heikkinen. Solving the shape of the Earth by using digital density models. Report 81:2, Finnish Geodetic Institute, Helsinki, 1981.

Weikko A. Heiskanen and Helmut Moritz. *Physical Geodesy*. W.H. Freeman and Company, San Francisco, 1967.

Reino A. Hirvonen. *Matemaattinen geodesia*. Teknillisen Korkeakoulun Ylioppilaskunta, Otaniemi, 1972.

Pasi Häkli, Jyrki Puupponen, Hannu Koivula, and Markku Poutanen. Suomen geodeettiset koordinaatistot ja niiden väliset muunnokset. Research note (in Finnish) 30, Finnish Geodetic Institute, Masala, 2009. URL http://www.fgi.fi/fgi/sites/default/files/publications/gltiedote/GLtiedote30.pdf.

IERS. IERS Conventions (1996). (IERS Technical Note No. 21). Technical report, Central Bureau of IERS - Observatoire de Paris, 1996. URL http://www.iers.org/nn_11216/IERS/EN/Publications/TechnicalNotes/tn21.html.

IERS. IERS Conventions 2003. IERS Technical Note 32. Technical report, Bundesamt für Kartographie und Geodäsie, Frankfurt am Main, 2003. URL http://www.iers.org/nn_11216/IERS/EN/Publications/TechnicalNotes/tn32.html.

IERS, ICRS Centre, maintained. URL https://www.iers.org/IERS/EN/Organization/ProductCentres/ICRSCentre/icrs.html. Accessed March 24, 2019.

ITRF, Transformation Parameters, maintained, Institut Géographique National, France. URL http://itrf.ensg.ign.fr/trans_para.php. Accessed March 23, 2019.

ITRF, Transformation Parameters between ITRF2005 and ITRF2008, maintained, Institut Géographique National, France. URL http://itrf.ensg.ign.fr/ITRF_solutions/2008/tp_08-05.php. Accessed March 23, 2019.

JPL, Solar System Dynamics, maintained. URL https://ssd.jpl.nasa.gov/. Accessed March 24, 2019.

JUHTA. JHS196. EUREF-FIN coordinates in Finland. Technical report, Advisory Committee on Information Management in Public Administration, 2016a. URL http://www.jhs-suositukset.fi/suomi/jhs196. Accessed March 9, 2019.

JUHTA. JHS197. EUREF-FIN coordinate systems, related conversions and map sheet distribution. Technical report, Advisory Committee on Information Management in Public Administration, 2016b. URL http://www.jhs-suositukset.fi/suomi/jhs197. Accessed March 9, 2019.

Juhani Kakkuri and Lasse Kivioja. Global positioning: The early Finnish connection. *Physics Today*, 65(11):8, 2012. doi: 10.1063/PT.3.1763. URL http://www.physicstoday.org/resource/1/phtoad/v65/i11/p8_s1?bypassSSO=1.

Daniel Kehlmann. *Measuring the World*. Rowolt Verlag, 2006. Original "Die Vermessung der Welt," 2005, translated Carol Brown Janeway.

Felix Klein and Arnold Sommerfeld. *Über die Theorie des Kreisels.* B.G. Teubner, Leipzig, 1897. URL https://archive.org/details/ berdietheoriede02sommgoog.

Arthur Koestler. *The Watershed: A Biography of Johannes Kepler.* Doubleday, 1960. Abridged version of *The Sleepwalkers.*

Arthur Koestler. *The Sleepwalkers: A History of Man's Changing Vision of the Universe.* Danube edition. Macmillan, 1968. URL http://books.google.fi/ books?id=csOnAAAAYAAJ.

Paul Edwin Kustaanheimo and Edward L. Stiefel. Perturbation theory of Kepler motion based on spinor regularization. *Journal für die reine und angewandte Mathematik,* 1965(218):204–219, 1965. doi: 10.1515/crll.1965.218. 204. URL https://www.degruyter.com/view/j/crll.1965.issue-218/crll. 1965.218.204/crll.1965.218.204.xml.

Pierre-Simon Laplace. *Traité de Mécanique Céleste.* Imprimerie de Crapelet, Paris, 1799. URL http://archive.org/stream/traitdemcani02lapl#page/146/mode/ 2up. Tome II, Livre III.

John M. Lee. *Riemannian Manifolds - An Introduction to Curvature.* Springer, New York, 1997.

Joseph Liouville. Extension au cas des trois dimensions de la question du tracé géographique, Note VI, pp. 609–617. In *G. Monge: Applications de l'analyse à la géométrie.* Bachelier, Paris, 1850. URL http://www.archive.org/stream/ applicationdela00unkngoog#page/n630/mode/2up.

Michael E. Mann. *The Hockey Stick and the Climate Wars: Dispatches from the Front Lines.* Columbia University Press, 2012. ISBN 9780231152549.

Donald E. Marshall. Numerical conformal mapping software: zipper, maintained. URL https://sites.math.washington.edu/~marshall/zipper.html. Accessed March 23, 2019.

Mark Monmonier. *How to lie with maps.* University of Chicago Press, 1996. 2nd edition.

Helmut Moritz. Geodetic Reference System 1980. In Carl Christian Tscherning, editor, *Geodesist's Handbook 1992,* pages 187–192. International Association of Geodesy, Copenhagen, 1992.

NASA, Joukowski Transformation, maintained, Glenn Research Center. URL https: //www.grc.nasa.gov/www/k-12/airplane/map.html. Accessed April 1, 2019.

Rolf Nevanlinna. On differentiable mappings. In L. Ahlfors et al., editor, *Analytic functions,* pages 3–9. Princeton University Press, 1960.

Aimo Niemi, editor. *Yrjö Väisälä – Tuorlan taikuri.* Fysiikan kustannus, 1991. ISBN 951-96117-3-8.

NIMA. NIMA Technical Report TR8350.2, "Department of Defense World Geodetic System 1984, Its Definition and Relationships With Local Geodetic Systems". Technical report, National Imagery and Mapping Agency, 3 January 2000. URL http://earth-info.nga.mil/GandG/publications/tr8350.2/wgs84fin.pdf.

NLS. Struve Geodetic Arc in Finland, maintained. URL https://www.maanmittauslaitos.fi/en/about-nls/themes/struve-geodetic-arc. Accessed March 23, 2019.

Matti Ollikainen. GPS-koordinaattien muuntaminen Kartastokoordinaateiksi. Research note (in Finnish) 8, Geodeettinen laitos, Helsinki, 1993.

Matti Ollikainen, Hannu Koivula, and Markku Poutanen. The densification of the EUREF network in Finland. Publication 129, Finnish Geodetic Institute, Masala, Finland, 2000.

Teuvo Parm. Kansallisen koordinaattijärjestelmän luominen Suomessa. *Maanmittaus,* 63(1), 1988.

Rich Pawlowicz. M_Map: a mapping package for MATLAB, maintained. URL www.eoas.ubc.ca/~rich/map.html. Accessed March 23, 2019.

Arno Peters. *Die Neue Kartographie/The New Cartography.* Friendship Press, New York, 1983.

Yuri G. Reshetnyak. Liouville's theorem on conformal mappings for minimal regularity assumptions. *Siberian Mathematical Journal,* pages 631–634, 1967. transl. from Sibirskii Mathematicheskii Zhurnal 8 (1967), 835–840.

V. Frederick Rickey and Philip M. Tuchinsky. An Application of Geography to Mathematics: History of the Integral of the Secant. *Mathematics Magazine,* 53(3), May 1980. URL http://www.maa.org/mathdl/mm/0025570x.di021115.02p0115x.pdf.

SeeSat-L, Visual Satellite Observing FAQ, archival site. URL http://www.satobs.org/faq/Chapter-09.txt. Accessed March 24, 2019.

Steve Silberman. *Neurotribes: the legacy of autism and the future of neurodiversity.* Penguin, New York, 2015.

Richard A. Snay. Introducing Two Spatial Reference Frames for Regions of the Pacific Ocean. *Surveying and Land Information Science,* 63(1):5–12, 2003.

Dava Sobel. *Longitude: the true story of a lone genius who solved the greatest scientific problem of his time.* Art House, 1995.

Kenneth Stephenson. *Introduction to Circle Packing: The Theory of Discrete Analytic Functions.* Cambridge University Press, Cambridge, 2005.

Wolfgang Torge. *Geodesy.* Walter de Gruyter, Berlin, New York, 2001. Third edition.

Carl Christian Tscherning and Knud Poder. Some geodetic applications of Clenshaw summation. *Bollettino di Geodesia e Scienze Affini,* 41(4):349–375, 1982. URL http://cct.gfy.ku.dk/publ_cct/cct80.pdf.

Aarne Veriö. Later phases and utilizing of the Northern part of the Struvean chain. *Geodeet,* 6(30):27–30, 1994. URL http://www.aai.ee/muuseum/Reprints/HTML/index.html?struvetriangulatsiooniahela.htm.

Martin Vermeer, Mauri Väisänen, and Jari Mäkynen. Paikalliset koordinaatistot ja muunnokset. Julkaisuja 37, Teknillisen korkeakoulun Geodesian laboratorio, Otaniemi, 2004.

Yrjö Väisälä. Maan toinen kuu. *Tähtitaivas,* (6):10–13, 1946. URL https://scienca-revuo.info/article/download/334/313. Version in Esperanto, translation Liisi Oterma.

Paul Wessel and Walter H.F. Smith. Free software helps map and display data. *EOS Transactions of the American Geophyscal Union,* 72(3):293–305, 1990.

Wikimedia Commons, Parallel ruler. URL https://commons.wikimedia.org/wiki/File:FieldsParallelRule.jpg. Accessed 2019-06-14.

Wikimedia Commons, Riemann, portrait. URL https://commons.wikimedia.org/wiki/File:Georg_Friedrich_Bernhard_Riemann.jpeg. 1863. Accessed 2019-06-14.

Wikipedia, Broom Bridge. URL http://en.wikipedia.org/wiki/Broom_Bridge. Accessed March 23, 2019.

Wikipedia, Gall-Peters projection. URL https://en.wikipedia.org/wiki/Gall%E2%80%93Peters_projection. Accessed March 23, 2019.

Wikipedia, Gravity Probe B. URL https://en.wikipedia.org/wiki/Gravity_Probe_B. Accessed April 3, 2019.

Wikipedia, Levi-Civita symbol. URL https://en.wikipedia.org/wiki/Levi-Civita_symbol. Accessed March 23, 2019.

Wikipedia, Mercator 1569 world map. URL https://en.wikipedia.org/wiki/Mercator_1569_world_map. Accessed March 23, 2019.

Wikipedia, Prague bell tower. URL https://en.wikipedia.org/wiki/File:Schema_Orloj_pragueorlojhzenilc.jpg. Accessed March 23, 2019, © 2007 Hector Zenil (CC BY-SA 2.5).

Wikipedia, Shape operator. URL https://en.wikipedia.org/wiki/Differential_geometry_of_surfaces#Shape_operator. Accessed March 23, 2019.

Wikipedia, Sophie Germain. URL https://en.wikipedia.org/wiki/Sophie_Germain. Accessed March 23, 2019.

Wikipedia, Sturm–Liouville theory. URL https://en.wikipedia.org/wiki/Sturm%E2%80%93Liouville_theory. Accessed March 23, 2019.

Wikipedia, Taylor's theorem. URL https://en.wikipedia.org/wiki/Taylor%27s_theorem. Accessed March 23, 2019.

Wikipedia, William Sealy Gosset. URL http://en.wikipedia.org/wiki/William_Sealy_Gosset. Accessed March 23, 2019.

Sarah Williams. The Old Astronomer, 1868. URL http://adsabs.harvard.edu/full/1933PA.....41..235.

Helmut Wolf. The Helmert block method, its origin and development. In *Proceedings of the Second International Symposium on Problems Related to the Redefinition of North American Geodetic Networks*, pages 319–326, 1978.

ZAH, Apparent Places, maintained University of Heidelberg, Zentrum für Astronomie. URL http://wwwadd.zah.uni-heidelberg.de/datenbanken/ariapfs/index.php.en. Accessed March 24, 2019.

Index

Printed in the United States
by Baker & Taylor Publisher Services